AF689828

Numerical methods for the personal computer

PRENTICE-HALL, INC., Englewood Cliffs, New Jersey 07632

TERRY E. SHOUP

Texas A&M University

Numerical methods for the personal computer

Library of Congress Cataloging in Publication Data

Shoup, Terry E. (date)
Numerical methods for the personal computer.

Includes index.
1. Numerical analysis—Data processing
2. Microcomputers. I. Title.
QA297.S476 1983 519.4'028'542 83-3258
ISBN 0-13-627208-8

Editorial/production supervision: *Lynn Frankel*
Cover design: *Diane Saxe*
Manufacturing buyer: *Gordon Osbourne*

© 1983 by Prentice-Hall, Inc., Englewood Cliffs, New Jersey 07632

All rights reserved. No part of this book
may be reproduced, in any form or by any means,
without permission in writing from the publisher.

Printed in the United States of America

10 9 8 7 6 5 4 3 2 1

ISBN 0-13-627208-8

Prentice-Hall International, Inc., *London*
Prentice-Hall of Australia Pty. Limited, *Sydney*
Editora Prentice-Hall do Brasil, Ltda., *Rio de Janeiro*
Prentice-Hall Canada Inc., *Toronto*
Prentice-Hall of India Private Limited, *New Delhi*
Prentice-Hall of Japan, Inc., *Tokyo*
Prentice-Hall of Southeast Asia Pte. Ltd., *Singapore*
Whitehall Books Limited, *Wellington, New Zealand*

*To the memory of Oscar L. Shoup and David R. Spoon,
who helped me learn about inventiveness, electronics,
and a host of other fun things.*

Contents

5 ORDINARY DIFFERENTIAL EQUATIONS

6 NUMERICAL INTERPOLATION AND CURVE FITTING

7 NUMERICAL DIFFERENTIATION AND INTEGRATION

APPENDIXES

The small computer is a truly remarkable tool for digital manipulation. Recent growth trends in the area of microcomputer technology predict the widespread use of small computers in many aspects of our lives in the decades ahead. Yet with all its widespread acceptance in the public domain, the small computer has seen relatively little use in applications for numerical methods in science and engineering. A primary reason for this fact is that most numerical methods texts and software packages are written from the perspective of the large, mainframe computer. Yet the use of the small computer as a tool for implementing certain types of numerical methods problems is rapidly becoming preferable to the larger computer for reasons of cost, utility, and convenience.

Of the limited information and software that is available for implementing numerical methods on small computers, very little has been developed by those who have an appreciation for both the power of modern numerical methods and the versatility of the small computer. It is the purpose of this text to exploit the best characteristics of these combined domains. This text is intended to provide not only a source of information about the area of numerical methods, but also to provide a software data base containing quality algorithms tailored for implementation on the small computer.

The text and its accompanying software should provide a package suitable for teaching numerical methods courses at universities and for individual scientific and engineering personnel who want to expand the utility of their computational equipment.

Preface

The spectrum of topics in this text includes the problem areas most frequently encountered in scientific and engineering problem solving. The basic approach of this text is one in which fundamental topics in numerical methods are presented along with the key mathematical relationships necessary for the development of useful algorithms. Then special-purpose software is provided to implement the algorithms on the small computer. Example applications are presented. A summary is presented at the end of each chapter to help the user choose the best algorithm for a given problem-solving task and to alert the user to potential problems in the application of these algorithms on the small computer. End of chapter references are provided to help the user discover more information about specific algorithms. For those who find the BASIC programs used in this text to be useful, a software package will soon be available.

The author would like to thank all those who contributed to the construction of this text and the approach it provides. Special thanks for encouragement are due to Dr. R.H. Page. A special note of appreciation is due to those of my colleagues who provided helpful suggestions to improve the text. The author is especially grateful to Dr. Don Riley, Dr. Farrokh Mistree, Dr. Douglas Green, and Dr. Ken Waldron. A special thank you is in order for Billie Gresham, whose remarkable talents and cheerful disposition made the preparation of the draft manuscript materials a pleasant activity. Finally, I want to express my gratitude to my family for their patience and encouragement during the many hours spent on this project.

TERRY E. SHOUP

Numerical methods for the personal computer

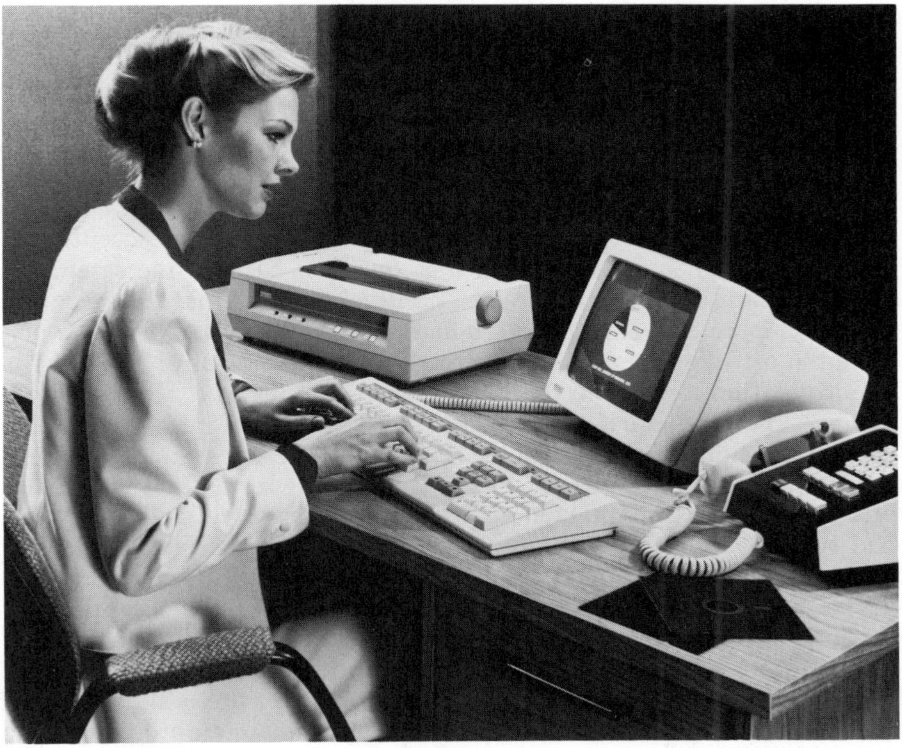

Because of its remarkable capability and size, the microcomputer is rapidly gaining widespread use in scientific and engineering problem solving. (*Photo courtesy of Digital Equipment Corporation.*)

Introduction

1

We live in an era of unprecedented scientific and engineering progress. In observing the technological advancements of the past few decades and how they have influenced the way we presently live and work, two exciting trends become apparent. The first of these is that significant achievements have tended to occur with increasing frequency as time progresses. The second trend is that the rate at which we accept these achievements and integrate them into our lives has also increased. As an illustration of these two trends, consider the fact that the invention of the telephone by Alexander Graham Bell in 1876 was slow to be perfected and to be accepted commercially. It was not until 1954 that a majority of households in this country had equipment for long-distance dialing. Yet the equally significant development of the first modern home television set in 1939 took less than 10 years to gain widespread acceptance. In 1974, the U.S. Census Bureau reported that 97 percent of U.S. households contained at least one television set and 45 percent had two or more sets. The rationale for this contrast seems quite clear. Early technological advancements in communications and product-delivery systems actually created a situation in which the acceptance of the later invention was accelerated. Thus, expanded technologies have a synergistic effect in fostering further innovation. The phenomenon is even more pronounced in our present era. Perhaps the best example of this is the microcomputer revolution. The first microprocessor chip was developed in 1971 and the first microcomputer appeared in 1975. Less than a decade later there are nearly a half-million microcomputers in use for tasks ranging from entertainment to business and scientific applications. Recent industrial surveys predict that nearly every home in this nation will have a microcomputer before the end of the decade. The microcomputer may well be as influential to our future life-style as was the development of the telephone or the television in the past. Contemporary writers have predicted a significant role for the microcomputer in the next social revolution following the industrial revolution.

Application areas for the microcomputer continue to emerge as the computational capability and hardware versatility of this device expand. Tasks once thought suitable for only large mainframe computers are now not only physically practical but are, in fact, economically preferably for implementation by a microcomputer. One such emerging application is the field of numerical problem solving in science and engineering (Figure 1-1). It is the purpose of this text to serve as a numerical methods resource to those who wish to apply the computational power of the microcomputer to scientific and engineering problem solving. This text will focus on three important goals:

1. To identify those characteristics of present and future microcomputer systems that suggest when to use and when not to use these important computational devices.

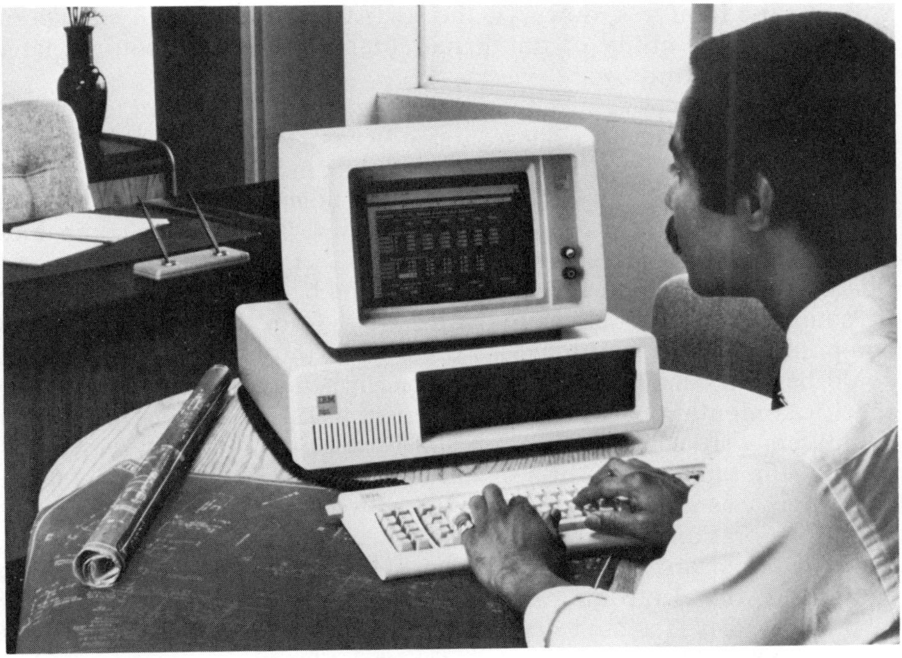

Figure 1-1 The microcomputer is rapidly becoming a significant tool for numerical problem solving in science and engineering. (*Photo courtesy of International Business Machines Corporation.*)

2. To identify those numerical tasks that are frequently encountered in engineering and scientific problem solving.
3. To present practical computational algorithms that represent an efficient merger of numerical method need with microcomputer capability.

1.1 THE DIGITAL COMPUTER

The first electronic digital computer was the ENIAC (from Electronic Numerical Integrator and Computer) built at the University of Pennsylvania between 1943 and 1949. Early computers such as this were large enough physically to fill a room the size of a small house. Yet, with all their size, their computational capability was rather primitive by present standards. The ENIAC computer contained over 18,000 vacuum tubes, and the overall reliability of this device was rather low owing to failures in these electronic tubes. Finding and replacing malfunctioning tubes took countless hours. The tubes also generated considerable heat and consumed large amounts of electric power. Even with its disadvantages, however, the ENIAC performed well enough to demonstrate the utility of the digital computer and to encourage later development of improved devices.

The transistor was invented in 1948 and by 1959 began to be used in digital computers to achieve a significant reduction in size, energy consumption, and costs, with a corresponding increase in both reliability and computational capability. The next quantum jump in size and capability came in the early 1960s when several transistor companies developed ways to place complete electronic circuits on the surface of silicon. These integrated circuits formed the basis for a new generation of computers with qualities much different from the early, large-scale computers used for multipurpose tasks. For the first time, computers could be made small enough and inexpensive enough to be dedicated to specific computational or data-management tasks. These small dedicated computers were called minicomputers because of their size, which was roughly the same as that of a small file cabinet. The first of these minicomputers was the PDP-8 manufactured in 1965 by the Digital Equipment Corporation. During this same time period, the integrated circuit made an important impact on another field of computational electronic equipment, the electronic calculator. In the late 1960s the capabilities of these devices increased dramatically, while the purchase price fell by more than an order of magnitude. The credit for this economic paradox is due mainly to improved manufacturing technology. It is not surprising then that the next breakthrough in computational equipment came on the interface between the calculator and the digital computer. Like so many revolutionary inventions, this innovation started with an attempt to solve a problem in one area and led to a breakthrough in another.

In 1971 the Intel Corporation was trying to design a single integrated circuit that would be a complete calculator on a single integrated-circuit chip. What resulted was a far more versatile device that we now call a microprocessor or a "computer on a chip." The technology that supported this important breakthrough is now called *large-scale integration* and allows the placement of thousands of transistors on a single silicon chip. The microprocessor is a complete computer central processing unit on a silicon chip smaller than a square centimeter (Figure 1-2). The microprocessor is capable of being used for a variety of dedicated applications including timing and controlling industrial processes, controlling traffic lights, guiding and controlling vehicles, and a host of other useful applications. One of the more interesting applications is possible when some memory is added to a microprocessor along with devices for input and output. Such a combination is called a *microcomputer.* It is interesting to note that, even though the size and cost of a microcomputer are small, it is easily capable of outperforming

Figure 1-2 The microprocessor "computer on a chip." (*Photo courtesy of Intel Corporation*.)

the large, expensive computers of the vacuum-tube and transistor eras. The classificational boundaries of definition for a large computer, a minicomputer, and a microcomputer are not clearly defined, owing to the continuous technological change that is contributing to their evolution. For convenience, the various types of computers presently in use may be classified in terms of size, speed, cost, and overall utility. Table 1-1 provides a comparison of characteristics for typical equipment in this rapidly evolving field. The implications of these characteristics to scientific and engineering problem solving will be discussed in the following paragraphs.

Size

At one time the physical size of a computer could be used as a measure of its overall computational capability; however, because of rapid changes in computer technology, this is no longer true. The same advancements make it more difficult to classify a computer as either large, medium, or small. Some minicomputers may be larger than some of the smaller of the large mainframe computers. In like manner, a microcomputer together with all its peripheral equipment may appear to be physically larger than some minicomputers. The significance of word size is that it indicates how many significant digits are possible in a given calculation. A typical 8-bit microcomputer is capable of calculations providing nine-significant-digit accuracy. In general, the more binary digits that are used in a computer word, the higher will be the number of significant digits of accuracy. Double precision capability

Table 1-1 Comparison of Characteristics for Various Types of Computers

Characteristic	*Medium/Large Computer*	*Minicomputer*	*Microcomputer*
Size			
Physical size	Room sized	Desk sized	Typewriter sized
Word size	32–64 binary digits	16–32 binary digits	8–16 binary digits
Max. memory size	12,000 K bytes	4000 K bytes	128 K bytes
Speed			
CPU cycle time	<70 nanoseconds	~100 nanoseconds	>200 nanoseconds
Memory cycle time	<250 nanoseconds	~300 nanoseconds	>400 nanoseconds
Cost	\$2 to \$4 million	\$20 to \$40,000	\$2 to \$4000
Utility			
Operating system	Multiprogramming (large numbers)	Multiprogramming (a few)	Single task
Languages supported	Most commonly used languages	Some	A few
Operating requirements	Special space and trained operator	Limited need for an operator	User operated
Available applications packages	Excellent	Good	Limited at present

frequently used by larger computers to expand accuracy at the expense of storage space is seldom available in microcomputers. Comparisons based on the size of working memory can be misleading. This number should be viewed in terms of word size and in terms of how much of the maximum memory size is actually available to a single user at a given time. As peripheral storage equipment becomes faster in access time, the limitations imposed by an active memory size of 64 K bytes are less severe. In general, it is rare for an engineering or scientific calculation to use all the memory capacity in a large computer. Such applications require the use of the large mainframe computer and are the type of task for which it is most efficiently suited. The microcomputer is best suited for computational applications requiring modest memory storage capacity and modest accuracy.

Speed

In spite of its small size, the microcomputer is surprisingly fast in operation. Based on CPU cycle time, the microcomputer operates at a rate of about one-tenth of the speed of its larger counterpart. Since many engineering and scientific problems performed on a large mainframe computer take less than 6 seconds of CPU time, it would seem reasonable to predict that programs of this same level of complexity would execute in less than 1 minute on a microcomputer. This rule of thumb should, of course, be applied with care since most engineering and scientific programs require more time for I/O operations than for actual CPU time. For these situations the processing speed is limited more by the speed of the peripheral units than by CPU speed. Since most microcomputers are connected to CRT screens or to slow-speed printers, it would seem prudent to restrict their use to problems requiring moderate amounts of overall processing time and moderate amounts of output.

Cost

The cost justification is perhaps one of the strongest arguments for using a microcomputer to do engineering and scientific problem solving. The normalized purchase price based on memory size for a microcomputer is about 5¢ per byte and for a large mainframe computer about 40¢ per byte. This eightfold advantage is possible because the microcomputer is designed to perform only one task at a time. A similar argument based on normalized purchase price for processing speed is even more dramatic than that for storage capacity. The CPU processing time for a large computer is about an order of magnitude faster than that for a microcomputer, whereas the overall purchase price is three orders of magnitude greater. In addition, since the microcomputer does not require specially trained operators nor specially prepared operating environments, its overall operation cost may be far more favorable than that predicted from a comparison of purchase price. Because it is based on relatively new technologies, it is likely that the purchase price of microcomputer systems will continue to decrease in the future. This fact adds to the already strong economic reasons for using the microcomputer for computational tasks within its utility range.

Utility

A major difference between a large and a small computer is in the number of users that can simultaneously interact with the machine. Under the multiprogramming environment of a large computer, several hundred users can simultaneously use this computational resource. On the other hand, for the microcomputer, operating systems are designed so that the machine functions entirely for a single user. For this reason, microcomputers are often referred to as "personal" computers. In the area of computer languages supported, most large-scale computers support a wide variety of commonly used high-level languages. Microcomputers, on the other hand, frequently support only one higher-level language. In most cases this is the BASIC language. Although microcomputers can usually be programmed in machine language, this process is different from machine to machine and makes transporting software somewhat difficult. Most scientific and engineering problem solving is presently accomplished using the FORTRAN language. Because of the widespread popularity of the microcomputer, special versions of this language are becoming available. Nevertheless, there remains a strong need for numerical methods programs in the BASIC language. Over the past decade a number of excellent application packages have been developed in FORTRAN for use on a variety of large computers. Because it is such a new technological development, the microcomputer does not have a large number of such application packages for engineering and scientific use. There is every reason to believe that the availability of such packages will grow at a rapid rate in the near future.

Figure 1-3, from a recent issue of the *Wall Street Journal*, illustrates the projected growth rate for small computers in the United States. This trend clearly indicates that higher quantities of these useful devices will be available in future years. Figure 1-4, also from the *Wall Street Journal*, indicates that a limited number of firms presently supply a large percentage of the market demand for these devices. It seems clear that the expected future demand for small computers will foster many new entries into this rapidly growing field. In considering the use

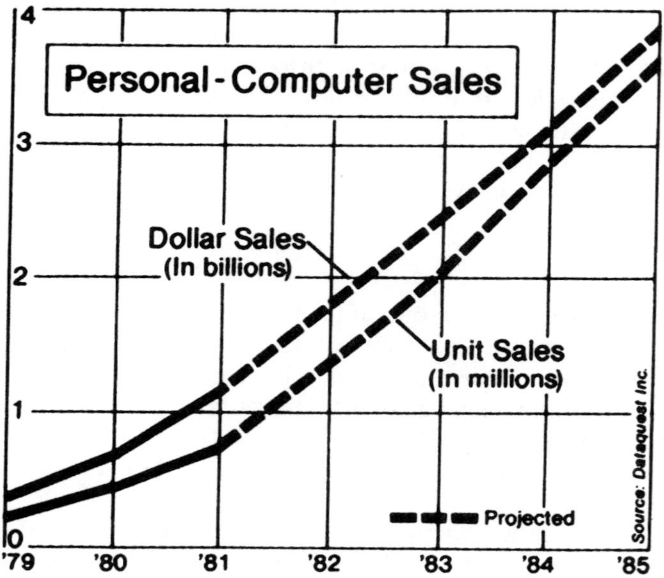

Figure 1-3 Projected growth rate for small computers. (*Reprinted by permission of the Wall Street Journal, © Dow Jones & Company, Inc., 1981. All rights reserved.*)

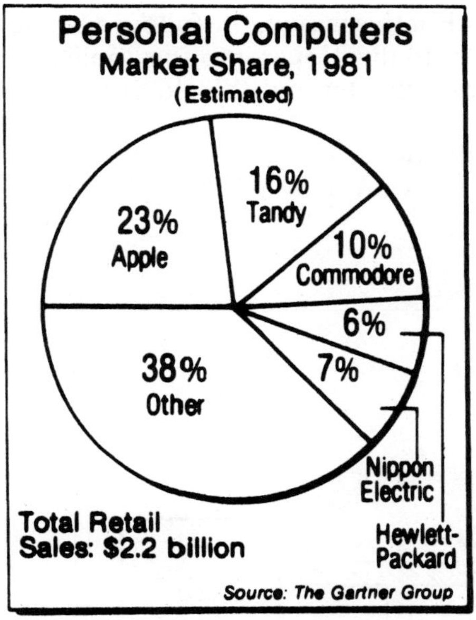

Figure 1-4 Primary suppliers of small computer equipment. (*Reprinted by permission of the Wall Street Journal, © Dow Jones & Company, Inc., 1981. All rights reserved.*)

of a small computer for scientific and engineering numerical problem solving, the following four characteristics should be kept in mind:

Time of run. Because the small computer runs at an order of magnitude slower than its larger counterpart, the small computer should be limited to applications that can be completed in a reasonable time.

Space. Because of the limited storage space and word size available through small computers, these devices should be used for solving problems that do not require large amounts of computational storage space.

Accuracy. Because most small computers operate with 8-bit words, applications requiring higher accuracy than available with this size of computation should be reserved for larger computers.

Amounts of I/O. Because of their relatively slow input and output capabilities, small computers should be utilized for problems requiring modest amounts of input and output information.

1.2 MICROCOMPUTER ARCHITECTURE

Although it is outside the scope of this text to present a rigorous treatment of the details of how a microcomputer operates, it is useful for the reader to understand the features that determine the performance characteristics of this unusual device. For this reason, the following material is presented. The user who wishes to dig deeper into this topic may find the references at the end of this chapter and the glossary in the appendix to be helpful.

Figure 1-5 shows the basic functional composition of a microprocessor central processing unit, also known as a CPU. The role of

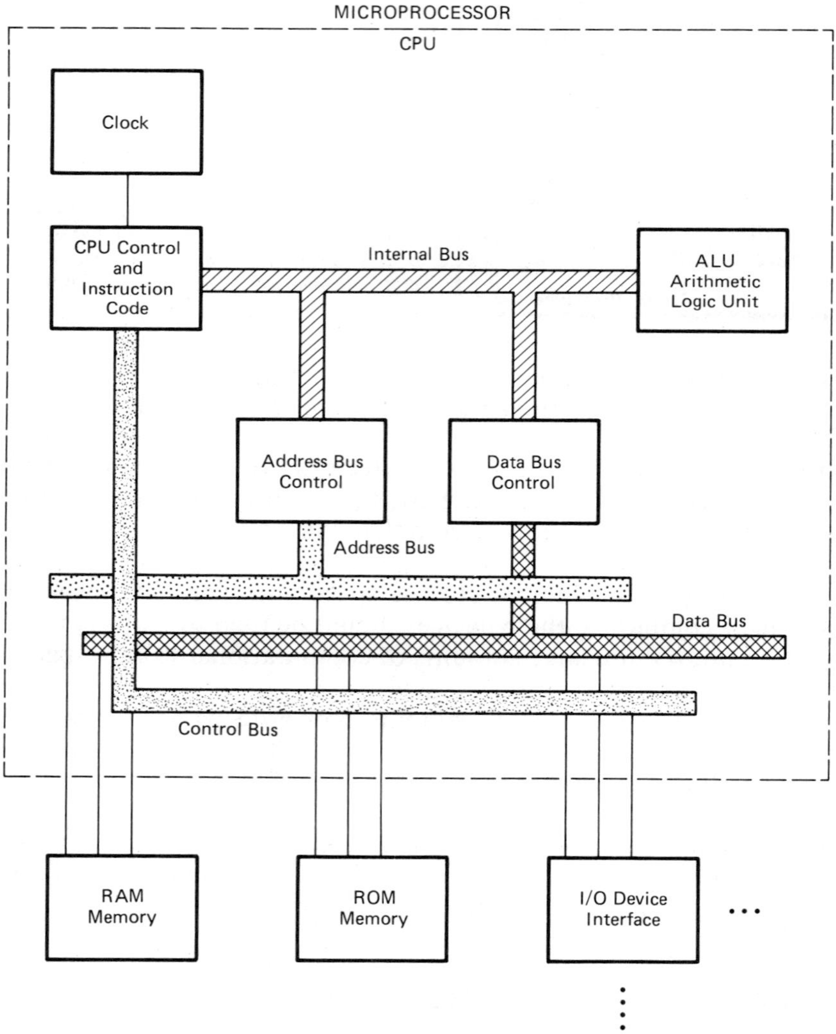

Figure 1-5 Microcomputer architecture.

each element contained in this diagram will be discussed in the following paragraphs.

Clock

The clock is a pulse generator capable of producing trains of evenly spaced pulses. Since the precise timing of functional events in a microcomputer is of extreme importance to its proper operation, the clock performs a vital role. The pulses are used to coordinate and synchronize the activity of all elements in the microcomputer. The speed of the clock will determine the overall speed of operation for the microcomputer up to the performance limitation of the system elements.

Arithmetic Logic Unit

The arithmetic logic unit (ALU) is the element in the microcomputer that performs all arithmetic and logical operations. In performing its task, the ALU makes use of various working registers that act as scratch pads for data manipulation.

Control Unit

The control unit is the control center of the entire microprocessor. Its job is to bring instructions in from the computer memory, interpret them, control their execution, and send the results to the appropriate places. The control unit has its own small memory containing the repertoire of instructions that are used by the microprocessor. In performing its tasks, the control unit also manages such housekeeping chores as holding the address of the main memory location from which the next instruction will be taken and setting flags and condition codes to indicate the current state of CPU activity. Several clock pulses are necessary to handle an instruction. This is because an instruction must be fetched from memory, decoded, and finally executed. The fetch, decode, and execution operations each involve one or more clock pulses called a *machine cycle*, and the entire sequence of operations is called an *instruction cycle*.

Buses

In the microprocessor the information connections between functional elements are called buses. The internal bus is used to pass information between internal elements such as the control unit and the ALU. The *data bus* allows information to flow in either direction so that the microprocessor can communicate with external elements. The *address bus* is a one-way line of information flow that sends an address from the microprocessor to external elements such as memory or I/O devices. The *control bus* is a communication line along which flows timing and status information. Some of these lines allow a flow of information to occur in two directions and some do not.

Most microcomputer manufacturers have developed their own printed circuit boards and bus configurations. Thus, in the early days of small computers, boards from one system were not compatible with those of another. Then in 1975 the MITS Corporation introduced the ALTAIR computer with a bus containing 100 pin connections. As the popularity of the ALTAIR grew, other manufacturers began producing products that were compatible with the ALTAIR bus. This bus, also known as the S-100 bus, has become the most widely used bus for small computer applications. Standards for the S-100 bus are now being written. It should be kept in mind, however, that this bus is not the only one presently used. Among the others in current use are the INTEL Multibus and the LSI-11 unibus of the Digital Equipment Corporation. The advantages of standard systems are clearly obvious to the user who wishes to join circuit boards from different manufacturers in order to capitalize on the best characteristics of diverse hardware.

Memory Units

The largest storage of information in a microcomputer is contained in its memory units. These units are divided into subunits called *registers*, each capable of holding one computer "word" consisting of a string of binary digits. Each register is uniquely identified by an *address*. In most microprocessors the words contain eight binary digits, and the 8-bit word is often given the name *byte*. For normal digital computer operation, there is a need for two different types of memory. The first type is used when it is desired that the contents of memory be permanent. This type of memory is called *read-only memory* (ROM) because it can only be read during normal computer operation. A frequently encountered use for ROM is to store a commonly used set of instructions such as a BASIC interpreter. Special types of ROM chips are also available. Among these are the PROM or *programmable read only memory*, which can be programmed only once, after which it becomes permanent, and the EPROM, *erasable programmable read only memory*, which can be reprogrammed again and again after being erased by ultraviolet light.

The second type of memory is used when the contents are to be frequently changed. This type of memory is called *random-access memory* (RAM). This type of memory can be read from and written into. RAM memory may be volatile (its contents are lost when the computer is powered down) or nonvolatile (its contents are not lost during power down).

Input/Output Devices and Interfaces

Because of its extreme versatility, a microprocessor may be connected to a variety of peripheral devices. Typical devices include CRT screens, keyboards, disk drive units, analog-to-digital converters, printers, and digitizing tablets. The interface circuits act as buffers to translate the information into the proper speed or form so that the microprocessor and the peripheral device can communicate correctly. This activity is necessary because humans and computers do not operate at the same speed. Another important application of an interface is to allow a computer to communicate with other computers to form a data communications network. To understand how interface devices work, it is necessary to recall that information stored in the small computer is contained in words consisting of 8 or 16 bits. If a whole word is to be transmitted at a single time, 8 or 16 transmission lines would therefore be required. When this is done, the interface is said to be *parallel operated*. Clearly, this mode of interface is efficient in terms of data speeds; however, it becomes cumbersome and expensive for long-distance transmission. This shortcoming suggests that transmission over a single line may be more realistic for certain applications. This type of interface is called *serial interfacing* and consists of the process of sending a single bit at a time. Since communication between different types of computers and peripheral devices is quite common, special standards for interfaces have been developed.

The IEEE 488 Standard has been widely accepted in instrumentation fields and is intended to be used to connect measuring instruments to a computer. This standard interface was developed by the Institute of Electrical and Electronics Engineers in 1975 and is now accepted as a standard by the American National Standards Institute (ANSI). Although the IEEE 488 Standard is often referred to as a "bus," it is not intended to be used to connect subsystems of a computer together as were the buses described previously.

The EIA RS-232C standard of the Electronics Industries Association is concerned with serial transmission of data between terminals and modems. It utilizes a special 25-pin connector, although for most applications only a few of these lines are actually used. A table of pin connections for the RS-232C interface is presented in Appendix E of this text.

Table 1-2 Common Numerical Tasks and Their Suitability for Microcomputer Solution

Numerical Method Topic	*Computational Time Required*	*Computational Space Required*	*Computational Accuracy Required*	*Amounts of Input and Output*	*Suitability for Microcomputer Solution*
Roots of algebraic and transcendental equations	Low	Low	Medium	Low	Good
Solution of simultaneous algebraic equations	Low	Medium	Medium	Low	Fair
Solution of eigenvalue problems	Medium	Medium	Medium	Medium	Fair
Curve fitting	Low	Low	Low	Low	Good
Interpolation and approximation	Low	Low	Low	Low	Good
Numerical integration and differentiation	Low	Low	Low	Low	Good
Ordinary differential equations	Low	Low	Low	Medium	Good
Partial differential equations	Medium	High	Medium	High	Poor
Optimization	High	Low	Medium	Low	Poor

1.3 THE MICROCOMPUTER AS A TOOL FOR NUMERICAL PROBLEM SOLVING

The nine categories of numerical problem-solving tasks listed in Table 1-2 are among the most frequently encountered problems in engineering and scientific analysis. These nine tasks are ranked according to the critical factors listed at the end of Section 1.2. From this table it becomes clear that the small computer is suited for reasonable applications in seven of these areas. These areas will be the topics of the remaining chapters of this text. Numerical algorithms will be provided using the BASIC language since this is the universal language of the small computer.

REFERENCES

1. BARDEN, W., *The Z-80 Microcomputer Handbook*, Howard W. Sams & Co., Inc., Indianapolis, Ind., 1978.
2. BUCHSBAUM, W.H., *Personal Computers Handbook*, Howard W. Sams & Co., Inc., Indianapolis, Ind., 1980.
3. FREIBERGER, S., and P. CHEW, *A Consumer's Guide to Personal Computing and Microcomputers*, Hayden Book Co., Inc., Rochelle Park, N.J., 1980.
4. GIVONE, D.D., and R.P. ROESSER, *Microprocessors/Microcomputers: An Introduction*, McGraw-Hill Book Co., New York, 1980.
5. KLINGMAN, E.E., *Microprocessor Systems Design*, Prentice-Hall, Inc., Englewood Cliffs, N.J., 1977.
6. LIBES, S., *Small Computer Systems Handbook*, Hayden Book Co., Inc., Rochelle Park, N.J., 1978.
7. McGLYNN, D.R., *Personal Computing*, John Wiley & Sons, Inc., New York, 1979.
8. SHOUP, T.E., *A Practical Guide to Computer Methods for Engineers*, Prentice-Hall, Inc., Englewood Cliffs, N.J., 1979.
9. SPENCER, D.D., *Computers and Programming Guide for Scientists and Engineers*, Howard W. Sams & Co., Inc., Indianapolis, Ind., 1980.

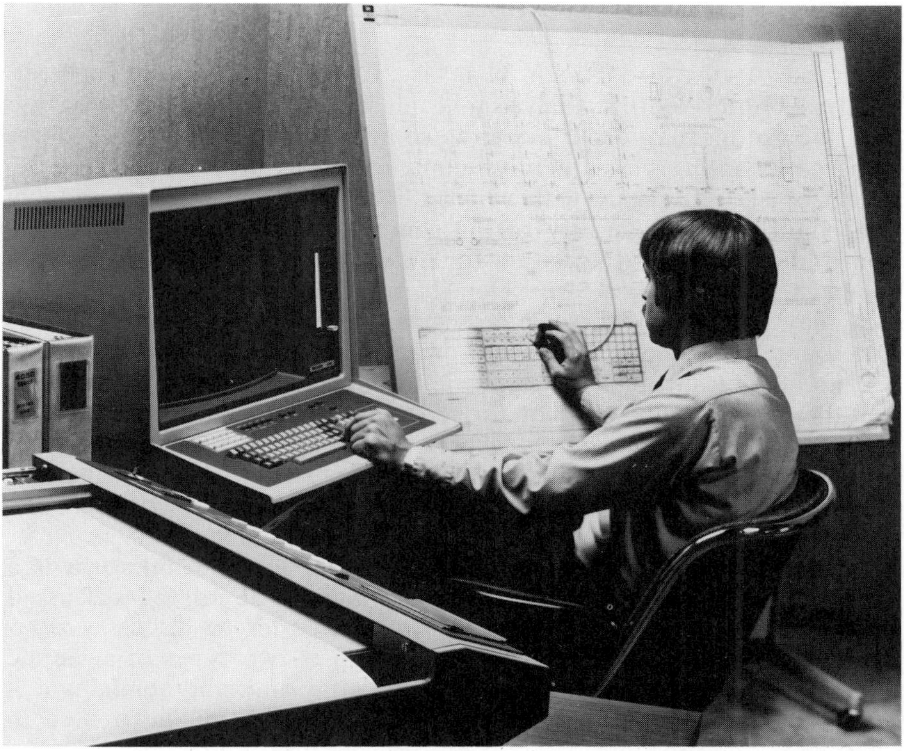

The microcomputer often forms the central part of a computer-aided design system. (*Photo courtesy of Tektronix*.)

Roots of algebraic and transcendental equations

2

The need to solve algebraic problems is a frequent situation in engineering design and scientific analysis. Problems of this type solved on the small computer may occur either as complete problems by themselves or as contributing parts to more complex procedures. In either case, the speed and efficiency with which the solution can be extracted will greatly influence the utility of the computational process. The selection of an appropriate algorithm for the solution of an algebraic system depends upon the class of problem being considered. Algebraic problems may be classified first according to the number of equations being solved and then according to the type and number of answers expected. A classification diagram for algebraic problems is shown in Figure 2-1. For the case of only one equation, the problem will be called *linear*, *polynomial*, or *transcendental* depending on whether it has one solution, *n* solutions, or an undetermined number of solutions. In the case of multiple equations, problems will be called *linear* or *nonlinear* depending on the mathematical nature of the equations involved.

The solution of a single, linear equation for one unknown is a sufficiently straightforward process so that it will not be discussed in this chapter. It is the purpose of this chapter to discuss various alternate methods for the solution of the first two types of algebraic systems, that is, the roots of transcendental and polynomial equations. The solution to multiple equation systems will be treated in Chapter 3.

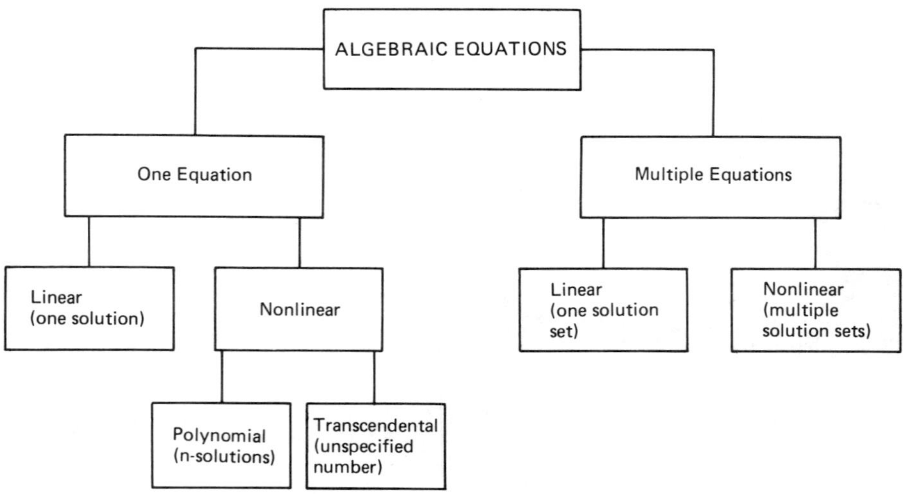

Figure 2-1 Classification of algebraic problems.

2.1 ROOTS OF A SINGLE NONLINEAR EQUATION

Two common classifications of nonlinear algebraic equations are the transcendental equation and the polynomial equation. Even though the techniques for the solution of these two categories are often the same, in this chapter they will be treated separately owing to the fact that polynomials possess special solution properties. We will consider the nonpolynomial equations first.

Nonlinear equations involving trigonometric relationships such as $\sin(x)$, $\cos(x)$, $\tan(x)$, or involving special functions such as $\log(x)$, e^x, and so on, are called transcendental. Solution methods for nonlinear equations of this type will be either direct or indirect. The direct methods find a solution by means of a set of formulas applied in a nonrepetitive fashion. The result is always an exact solution. A familiar example of a direct method is the quadratic formula technique for second-order polynomial equations. An indirect method, on the other hand, provides a process for achieving a solution as a result of the repeated application of an algorithm. The result is always approximate, although any desired degree of accuracy can usually be achieved. Iterative methods are best suited for computer implementation and thus will be discussed in detail in this chapter. In each of the methods presented, it will be assumed that the problem to be solved is that of finding the real roots (zeros) of the equation $f(x) = 0$. Although it is possible for nonpolynomial equations to have complex roots, the determination of complex roots is generally considered only for polynomial problems.

2.2 BINARY SEARCH METHOD

The rationale for the binary search method is based on the technique of interval halving. The logic flow diagram for this method is presented in Figure 2-2 and proceeds as follows. First, the function is evaluated at equally spaced intervals of x until two successive function values $f(x_n)$ and $f(x_{n+1})$ are of opposite signs. (The change in sign will indicate the existence of a root if the function is continuous.) On the range from x_n to x_{n+1}, the midpoint value is calculated using the formula

$$x_{\text{mid}} = \frac{x_{n+1} + x_n}{2}$$

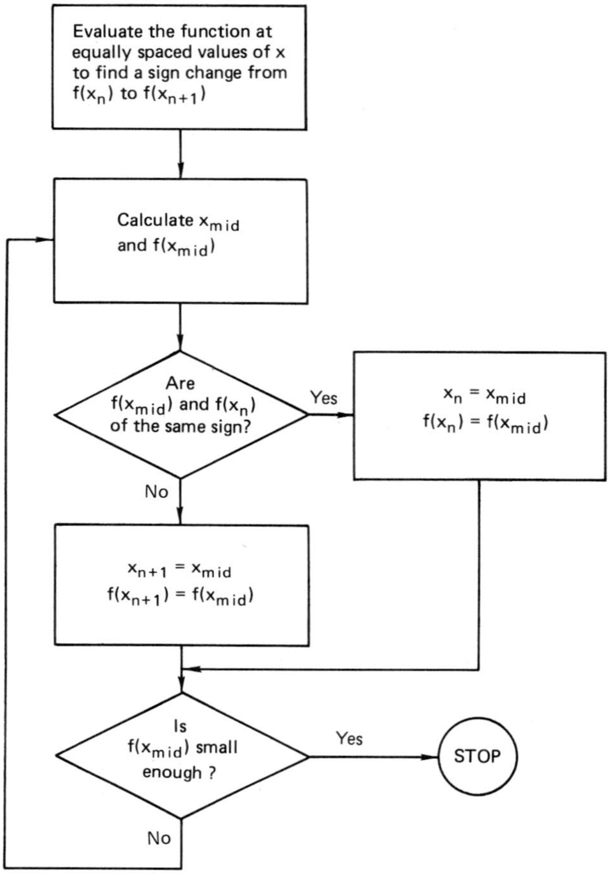

Figure 2-2 Logic flow diagram for the binary search method.

and the function value $f(x_{\text{mid}})$ is found. If the sign of $f(x_{\text{mid}})$ agrees with $f(x_n)$, it is used to replace $f(x_n)$; if not, it agrees with $f(x_{n+1})$ and replaces this value. The interval of uncertainty in which the root must lie is thus reduced. If the value of $f(x_{\text{mid}})$ is small enough, the process is terminated; otherwise the process is repeated. Figure 2-3 illustrates this procedure graphically. Although this method is not efficient in computational effort, it does give an improved approximation to the actual root with increasing number of evaluations. Once the first interval of uncertainty is identified, the interval of uncertainty can be reduced by a factor of 2^{-N} for N iterations.

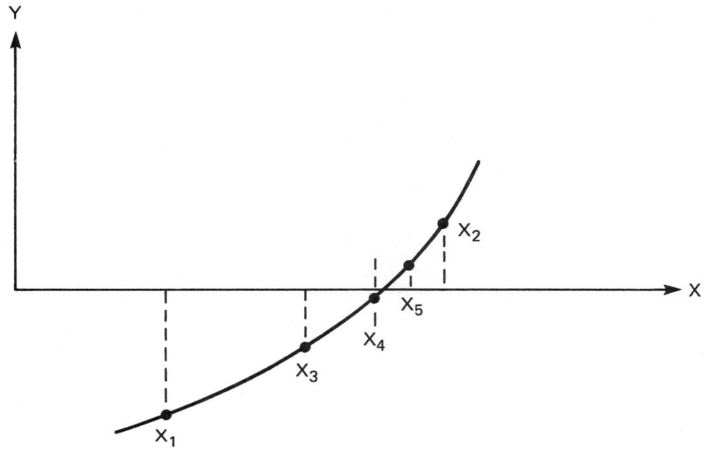

Figure 2-3 The binary search method.

2.3 FALSE POSITION METHOD

The method of false position is based on a linear interpolation between two values of the function that have opposite signs. This method often converges to a root more quickly than the binary search method. The logic flow diagram for this method is presented in Figure 2-4 and proceeds as follows. First the function is evaluated at equally spaced intervals of x until two successive functions values $f(x_n)$ and $f(x_{n+1})$ are found to have opposite signs. A line passing through these two points will have a root at

$$x^* = x_n - f(x_n)\,\frac{x_{n+1} - x_n}{f(x_{n+1}) - f(x_n)}$$

This value is used to find $f(x^*)$, which is in turn used to compare with $f(x_n)$ and $f(x_{n+1})$ to replace the one with similar sign. If $f(x^*)$ is not close enough to zero, the calculation process is repeated until the desired degree of convergence is achieved. Figure 2-5 illustrates this process graphically.

2.4 NEWTON'S METHOD

Newton's method of iteration is the most widely used of all iterative techniques. Its popularity is due to the fact that, unlike the previous

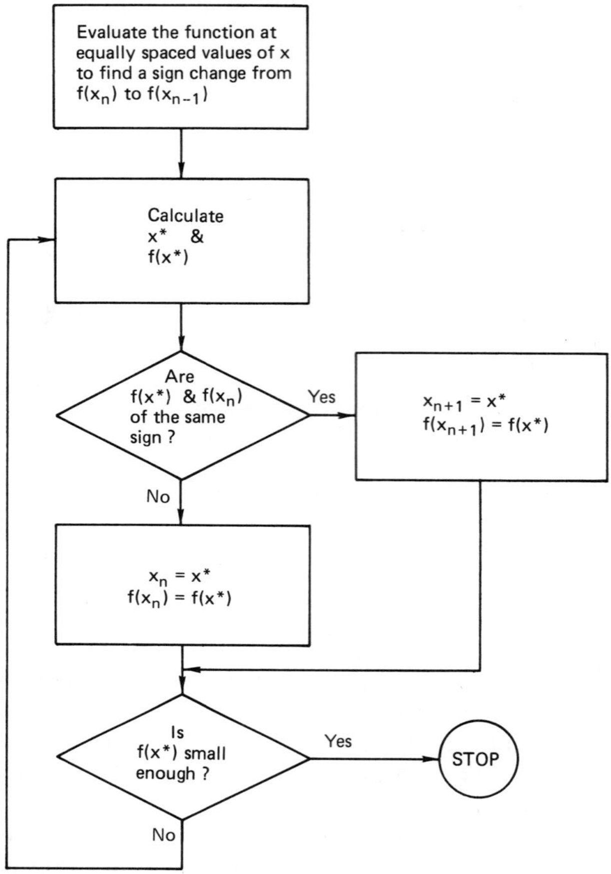

Figure 2-4 Logic flow diagram for the false position method.

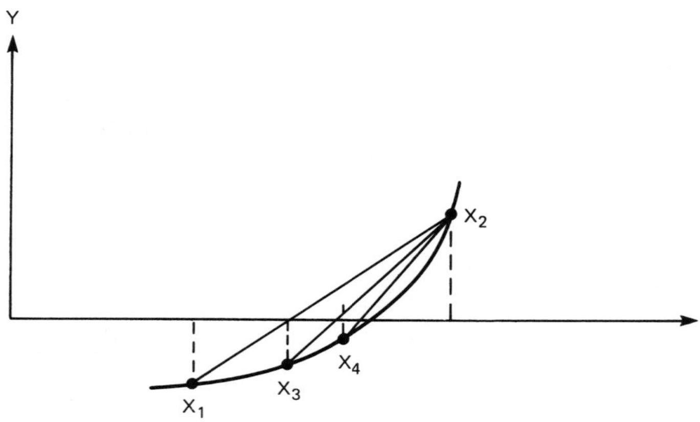

Figure 2-5 The method of false position.

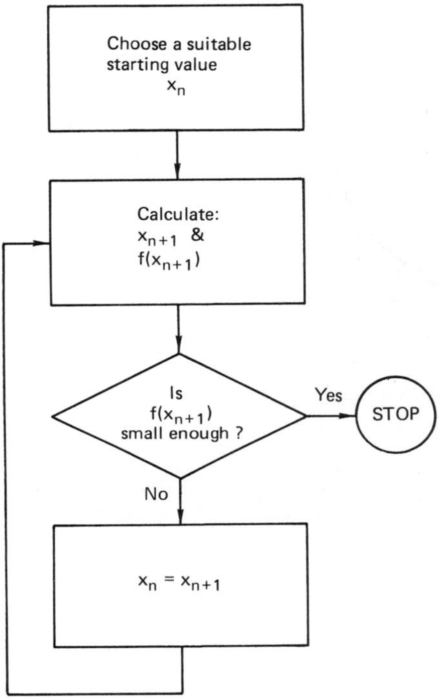

Figure 2-6 Logic flow diagram for Newton's method.

methods, it does not require the user to search for two function values of different sign in order to bracket the root. Rather than using an interpolation scheme based on two function values, this method uses extrapolation based on a line that is tangent to the curve at a point. The logic flow diagram for this method is shown in Figure 2-6. The method is developed from a Taylor's expansion of the form

$$x_{n+1} = x_n - \frac{f(x_n)}{f'(x_n)}$$

The value x_{n+1} is equivalent to the point where the curve tangent at x_n passes through the x axis. Since the curve $f(x)$ is likely not a line, the functional value $f(x_{n+1})$ is likely not exactly zero. For this reason the process is repeated using $x_n = x_{n+1}$ as a new base point. When the value of $f(x_{n+1})$ is sufficiently small, the process is terminated. Figure 2-7 illustrates Newton's method graphically. Clearly, the choice of location for the starting point will greatly influence the speed of convergence. If the slope of the curve $f'(x)$ goes to zero in the iterative process, the method has difficulty. In addition, it can be shown that if $f''(x)$ goes to infinity, the method will also fail to perform properly.

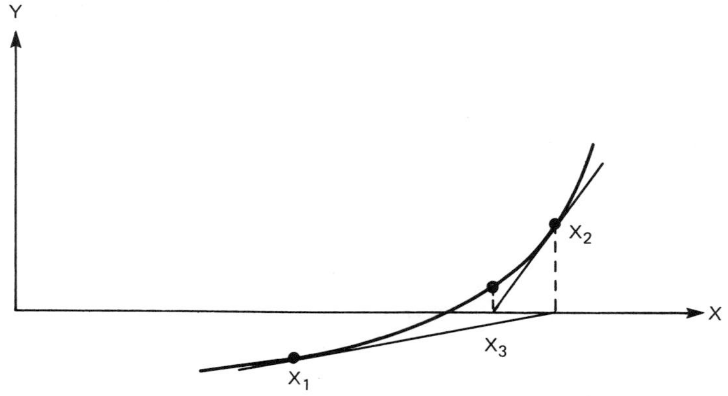

Figure 2-7 The Newton iteration method.

Since the conditions for a multiple root are $f(x) = 0$ and $f'(x) = 0$, Newton's method will not converge for this case. It should be noted that an alternative procedure sometimes used for checking convergence is a comparison of x_n and x_{n+1}.

2.5 SECANT METHOD

One disadvantage of Newton's method is that it requires the user to take a derivative of the function $f(x)$. In the event that it is inconvenient to find this derivative, an approximation may be used. This alternative is the basis for the secant method. If the derivative $f'(x_n)$ in the Newton method formula

$$x_{n+1} = x_n - \frac{f(x_n)}{f'(x_n)}$$

is replaced by means of two successive functional approximations in the formula

$$\text{slope}\ (x_n) = \frac{f(x_n) - f(x_{n-1})}{x_n - x_{n-1}}$$

the iteration formula becomes

$$x_{n+1} = x_n - \frac{f(x_n)}{\text{slope}\ (x_n)}$$

The logic flow diagram for this method will be the same as for Newton's method except that the iteration formula is slightly different. This method actually seeks the root by a combination of interpolation and extrapolation. Whenever it is operating in the interpolation mode, the method is equivalent to the method of false position. As with the Newton method, this technique may be terminated when consecutive values of x agree to within some acceptable error or when the function value $f(x)$ is acceptably close to zero. The secant method has the same convergence difficulties at a multiple root as does the method of Newton iteration.

2.6 DIRECT SUBSTITUTION METHOD

The direct substitution method is a straightforward technique that can be used if the function $f(x) = 0$ can be manipulated into the form

$$x = g(x)$$

Using this form, an iterative formula

$$x_{n+1} = g(x_n)$$

can be formulated. The method would proceed in the pattern indicated in Figure 2-8. Because of its basic simplicity, this method may

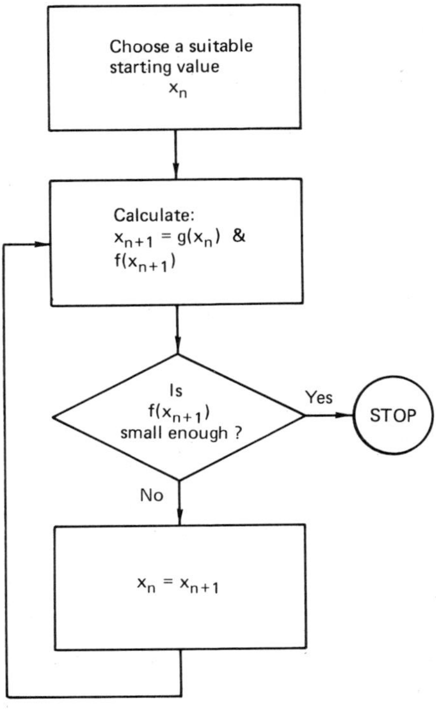

Figure 2-8 Logic flow diagram for the method of direct substitution.

appear to be attractive; however, the method does have convergence pitfalls. For this reason, any program using this algorithm should contain suitable checks to terminate the iterative process if it fails to converge.

EXAMPLE
2-1

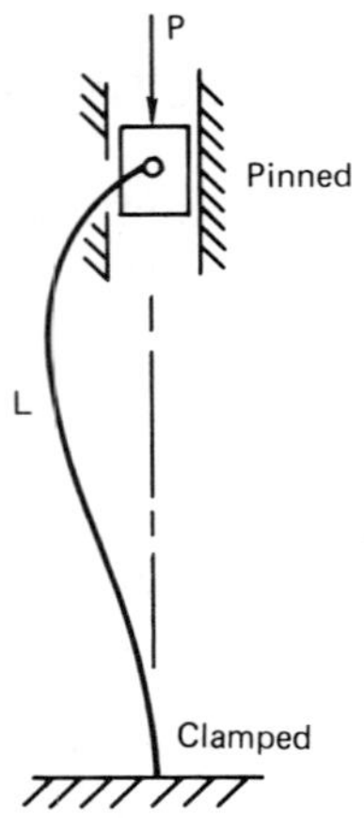

Suppose that it is desired to compute the buckling load for the clamped-pinned column shown in the figure. The algebraic relationship describing this critical load will be

$$\tan(x) = x$$

where

$$x = \sqrt{\frac{P}{EI}}\,L$$

and P = critical load for buckling
EI = flexural rigidity
L = length of the column

Thus the transcendental equation to be solved is

$$x - \tan(x) = 0$$

To find a solution to this problem, it is first necessary to determine approximately where a solution will lie. A simple sketch of $\tan(x)$ versus x can be used to show that the first nonzero root will lie in the third quadrant. To illustrate the relative computational advantages of different methods, two different root-solving algorithms will be used to solve this problem.

The first method to be used will be the method of false position. A BASIC computer program that implements this solution is given next. This program performs the root-solving task in two stages. In the first stage the search interval is divided into ten equal parts with a view toward finding a set of function values that have different signs. Once the sign change is found, the algorithm of Figure 2-4 is used to implement the second stage, in which the root value is further refined.

```
1000  REM **********************
1010  REM * THIS PROGRAM FINDS *
1020  REM * A ROOT TO THE      *
1030  REM * TRANSCENDENTAL EQN *
1040  REM *                    *
1050  REM *    0 = X - TAN(X)  *
1060  REM *                    *
1070  REM * ON THE RANGE       *
1080  REM *                    *
1090  REM *  PI < X < 1.5*PI   *
1100  REM *                    *
1110  REM * USING THE METHOD OF*
1120  REM * FALSE POSITION     *
1130  REM **********************
1140 :
1150  REM **********************
1160  REM * LOOK FOR A  CHANGE *
1170  REM * IN SIGN TO INDICATE*
1180  REM * THE PRESENCE OF A  *
1190  REM * ROOT.              *
1200  REM **********************
1210 :
1220 PI = 3.1415926
1230 XL = PI
1240 FL = XL -  SIN (XL) /  COS (
     XL)
1250 DX = 0.05 * PI
1260  FOR I = 1 TO 10
1270 XR = XL + DX
1280 FR = XR -  SIN (XR) /  COS (
     XR)
1290  IF FR * FL < 0 GOTO 1440
1300 XL = XR:FL = FR
1310  NEXT I
1320 :
1330  PRINT "NO SIGN CHANGE"
1340  END
1350 :
1360 :
1370  REM **********************
1380  REM * USE FALSE POSITION *
1390  REM * TO FIND IMPROVED   *
1400  REM * VALUE FOR ROOT     *
1410  REM * UNTIL ABS(F)<0.0001*
1420  REM **********************
1430 :
1440 XSTAR = XL - FL * (XR - XL) /
     (FR - FL)
1450 FS = XSTAR -  SIN (XSTAR) /
     COS (XSTAR)
1460  IF  ABS (FS) < 0.0001 GOTO
     1530
1470  IF FS * FL > 0 GOTO 1500
1480 XL = XSTAR:FL = FS
1490  GOTO 1440
1500 XR = XSTAR:FR = FS
1510  GOTO 1440
1520 :
1530  PRINT "ROOT IS ";XSTAR
1540  END
```

The output of this program is

```
ROOT IS 4.49341328
```

This program requires less than 2 seconds to run on an Apple II computer. The storage space required to implement this program could, of course, be reduced by eliminating the REM statements. One drawback to the method is that a first stage of the algorithm must find a sign change for the function in order to go on to the second stage. The second method to be tried does not have this disadvantage.

The second method to be used is the Newton iteration method. A BASIC computer program that implements this solution is given next. This program starts with an initial guess and proceeds to improve the value until the magnitude of the function is less than 0.0001.

```
1000  REM **********************
1010  REM * THIS PROGRAM FINDS *
1020  REM * A ROOT TO THE      *
1030  REM * TRANSCENDENTAL EQN *
1040  REM *                    *
1050  REM *   0 = X - TAN(X)   *
1060  REM *                    *
1070  REM * ON THE RANGE       *
1080  REM *                    *
1090  REM *  PI < X < 1.5*PI   *
1100  REM *                    *
1110  REM * USING THE METHOD OF*
1120  REM * NEWTON  ITERATION  *
1130  REM **********************
1140 :
1150 X = 4.3
1160  FOR I = 1 TO 30
1170 F = X -  SIN (X) /  COS (X)
1180 DF = 1. - 1. /  COS (X) ^ 2
1190 X = X - F / DF
1200  IF  ABS (F) < 0.0001 GOTO 1250
1210  NEXT I
1220  PRINT "NO SOLUTION AFTER 30 ITERATIONS"
1230  END
1240 :
1250  PRINT "ROOT IS ";X
1260  END
```

The output of this program is

```
ROOT IS 4.49340945
```

This program requires less than 2 seconds to run on an Apple II computer. The advantage to this method is that it is done in a single stage and thus requires far less lines to implement than the method of false position. The disadvantage to the method is that a poor first guess may lead to a lack of convergence on the part of the algorithm. For example, if starting values of $x = 4.0$ or $x = 5.0$ are tried, the method does not converge.

2.7 SOLUTION OF POLYNOMIAL EQUATIONS

Algebraic equations involving only the sum of integer powers of x are called *polynomial equations*. Their general form will be

$$x^n + a_0 x^{n-1} + \cdots + a_{n-2} x^1 + a_{n-1} = 0$$

Properties of Polynomials

Certain special properties about polynomial equations are useful in determining the nature of the solutions to these equations. These are as follows:

1. An nth order polynomial will have n roots. These roots may be real or complex.
2. If all the a_i coefficients are real, then all complex roots will appear in complex conjugate pairs.
3. The number of positive real roots is equal to or less than, by an integer, the number of sign changes in the a_i coefficients.
4. The number of negative real roots is equal to or less than, by an integer, the number of sign changes in the a_i coefficients if x is replaced by $-x$.

Although direct methods are available for the solution of second- and third-order polynomial equations, for higher-order equations indirect methods must be used. In the strict mathematical sense, once a root of a polynomial equation has been found by an indirect method, the root may be used to reduce the order of the polynomial by 1 by dividing by the linear factor (x - root). The result will be a polynomial of order $n - 1$. Although this procedure might seem attractive, it is not recommended because slight errors in the value of the initial root can cause the accumulation of substantial errors in the coefficients of the reduced polynomial. It should be noted that this procedure often does provide a means for selecting a reasonable guess for other roots once a few initial roots are already known.

The solution algorithms for transcendental equations presented in the previous sections can be used to find both real and complex roots of polynomials if the user is willing to utilize complex arithmetic. Unfortunately, small computers rarely support the use of complex variables and arithmetic. For this reason, methods based on real arithmetic that yield complex solutions are especially important to numerical analysis on the small computer.

2.8 LIN'S METHOD FOR COMPLEX ROOTS

A few special methods for finding complex roots are available. These nearly always involve a procedure for extracting a quadratic factor

$$x^2 + px + q$$

from the original polynomial. One common method of this type is Lin's method. It is based on the fact that a polynomial of the form

$$x^n + a_0 x^{n-1} + \cdots + a_{n-2}x + a_{n-1} = 0$$

can be written as

$$0 = (x^2 + px + q)(x^{n-2} + b_0 x^{n-3} + \cdots + b_{n-4}x + b_{n-3}) + Rx + S$$

In this expression $Rx + S$ is a linear remainder term that we desire to be zero. A zero remainder would mean that the original polynomial is exactly divisible by the quadratic factor. If like coefficients are compared between the two forms of the polynomial equation, the result will be

$$\begin{aligned}
b_0 &= a_0 - p \\
b_1 &= a_1 - pb_0 - q \\
&\vdots \\
b_i &= a_i - pb_{i-1} - qb_{i-2}, \qquad \text{for} \quad i = 2, 3, \ldots, n-3 \\
&\vdots \\
R &= a_{n-2} - pb_{n-3} - qb_{n-4} \\
S &= a_{n-1} - qb_{n-3}
\end{aligned}$$

The procedure for the iterative process that leads to a solution is illustrated in Figure 2-9. It proceeds as follows. First, initial guesses for p and q are made. These, together with the given a_i coefficients, are used to calculate b_0. The value of b_0 is, in turn, used to calculate b_1, and so on, until all coefficients down and through b_{n-3} have been calculated. If it is assumed that the remainder terms R and S are zero, the terms b_{n-4}, b_{n-3}, a_{n-2}, and a_{n-1} can be used in the last two equations to get improved values of p and q, say p^* and q^*. If the change in the p and q values is sufficiently small, the process is terminated. If the change is not small enough, the new values replace p and q and the process is repeated. As it turns out, this procedure amounts to the solution of two equations in two unknowns by direct iteration. This is a topic that will be covered in more depth in Chapter 3.

Another method based on the quadratic factor $x^2 + px + q$ uses Newton's method for the two equations and two unknowns. It is called Bairstow's method.

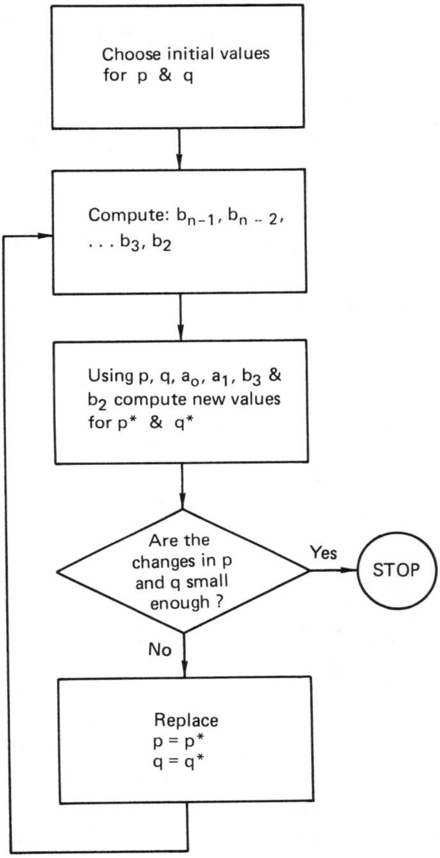

Figure 2-9 Logic flow diagram for Lin's method.

2.9 BAIRSTOW'S METHOD FOR FINDING THE ROOTS OF POLYNOMIALS

The basis for Bairstow's method is that initial guesses are made for the terms p and q in the quadratic factor presented in the previous formulation. These factors are then improved by means of correction factors:

$$p = p + \Delta p$$

$$q = q + \Delta q$$

In this process it is desired that the remainder terms $R(p, q)$ and $S(p, q)$ become zero. If these two functions are expanded in a Taylor's series expansion, the result will be

$$R(p + \Delta p, q + \Delta q) = R(p, q) + \frac{\partial R}{\partial p} \Delta p + \frac{\partial R}{\partial q} \Delta q + \text{higher-order terms}$$

$$S(p + \Delta p, q + \Delta q) = S(p, q) + \frac{\partial S}{\partial p} \Delta p + \frac{\partial S}{\partial q} \Delta q + \text{higher-order terms}$$

If one assumes that the correction factors move the selection of p and q toward zero remainder terms, then the left sides of these equations will be zero. If this is true and if the correction factors are small enough so that higher-order terms can be neglected, then the two equations in two unknowns can be solved to get

$$\Delta p = \frac{S \dfrac{\partial R}{\partial q} - R \dfrac{\partial S}{\partial q}}{\dfrac{\partial R}{\partial p} \dfrac{\partial S}{\partial q} - \dfrac{\partial S}{\partial p} \dfrac{\partial R}{\partial q}}$$

$$\Delta q = \frac{R \dfrac{\partial S}{\partial p} - S \dfrac{\partial R}{\partial p}}{\dfrac{\partial R}{\partial p} \dfrac{\partial S}{\partial q} - \dfrac{\partial S}{\partial p} \dfrac{\partial R}{\partial q}}$$

To find the partial derivatives of R and S with respect to p and q, it must be kept in mind that R and S are functions of the b_i values, which are, in turn, functions of p and q. Thus it becomes necessary to develop a sequence of partial derivative terms much like the b_i sequence developed previously. For example,

$$\frac{\partial b_0}{\partial p} = c_0 = -1$$

$$\frac{\partial b_1}{\partial p} = c_1 = -b_0 + p$$

$$\vdots$$

$$\frac{\partial b_i}{\partial p} = c_i = -b_{i-1} - pc_{i-1} - qc_{i-2}, \qquad \text{for} \quad i = 2, 3, \ldots, n-3$$

$$\vdots$$

$$\frac{\partial R}{\partial p} = -b_{n-3} - pc_{n-3} - qc_{n-4}$$

$$\frac{\partial S}{\partial p} = -qc_{n-3}$$

and

$$\frac{\partial b_0}{\partial q} = d_0 = 0$$

$$\frac{\partial b_1}{\partial q} = d_1 = -1$$

$$\vdots$$

$$\frac{\partial b_i}{\partial q} = d_i = -b_{i-1} - pd_{i-1} - qd_{i-2}, \qquad \text{for} \quad i = 2, 3, \ldots, n-3$$

$$\vdots$$

$$\frac{\partial R}{\partial q} = -b_{n-4} - pd_{n-3} - qd_{n-4}$$

$$\frac{\partial S}{\partial q} = -b_{n-3} - qd_{n-3}$$

The procedure for this method is shown in Figure 2-10. The Bairstow method keeps finding improvements to p and q in the form of correction factors until a desired solution is achieved. This may be deter-

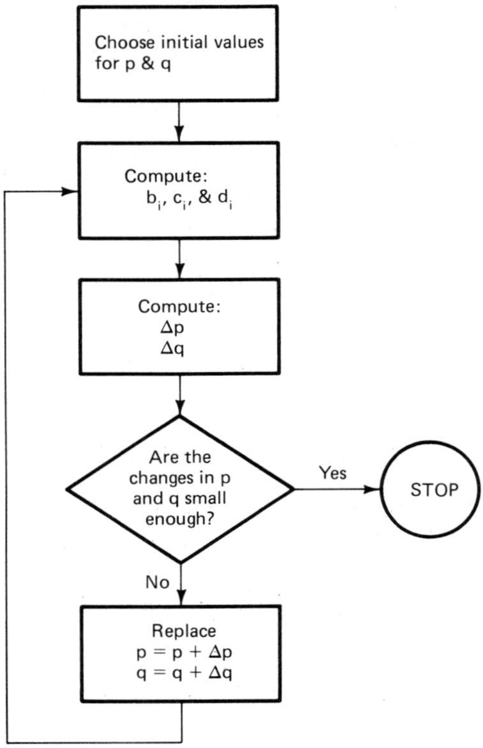

Figure 2-10 Logic Flow Diagram for Bairstow's Method.

mined by looking at how small R and S have become or by looking at how small p and q have become. This method is an improvement over Lin's method because it allows faster convergence and will sometimes converge when Lin's method will not. As it turns out, this method amounts to the solution of two equations in two unknowns by Newton–Raphson iteration. This topic will be covered in more depth in Chapter 3.

EXAMPLE 2-2 Suppose that it is desired to find all roots of the polynomial

$$x^5 + 4x^4 - 9x^3 + 14x^2 + 50x - 600 = 0$$

Since all roots are desired, Bairstow's method will be used. A BASIC program that finds these roots is now presented. This program finds first estimates for the roots from the reduced polynomial and then performs a final iteration cycle using the original polynomial. This program is actually designed to handle polynomials of odd or even order. If the polynomial is odd, there will be at least one real root. The iteration for this root uses a simple Newton iteration subroutine to find the root rather than a quadratic factor. In this program the polynomial coefficients are entered as elements of the array COF(N), the real parts of the roots found are stored in an array RR(N), and the imaginary parts of the roots found are stored in an array RI(N). The program also requires several working arrays A(N), B(N), C(N), D(N), PP(N/2 + 1), and QQ(N/2 + 1).

```
1000  REM ********************
1010  REM * THIS PROGRAM FINDS*
1020  REM * ALL ROOTS OF A    *
1030  REM * POLYNOMIAL WITH   *
1040  REM * REAL COEFFICIENTS *
1050  REM * BY BAIRSTOW'S     *
1060  REM * METHOD.           *
1070  REM ********************
1080 :
1090  DIM COF(5),A(5),B(5)
1100  DIM PP(3),QQ(3)
1110  DIM RR(5),RI(5)
1120  DIM C(5),D(5)
1130 :
1140 COF(0) = 4:COF(1) = - 9:COF
   (2) = 14
1150 COF(3) = 50:COF(4) = - 600
1160 N = 5
1170 :
1180  REM **APPLY BAIRSTOW'S
1190  REM **METHOD.
1200 :
1210  GOSUB 2000
1220 :
1230  REM **PRINT THE RESULTS
1240 :
1250  PRINT "THE ROOTS ARE AS FOL
   LOWS"
1260  PRINT "-------------------
   -------"
1270  PRINT "  REAL            IMAG
   INARY"
1280  PRINT "  PART               P
   ART"
1290  PRINT "-------------------
   -------"
```

```
1300  FOR I = 0 TO N - 1
1310  PRINT RR(I),RI(I)
1320  NEXT I
1330  PRINT "--------------------
-------"
1340  END
1350 :
1360 :
2000  REM **********************
2010  REM * THIS SUBROUTINE     *
2020  REM * APPLIES BAIRSTOW'S *
2030  REM * METHOD TO FIND ALL *
2040  REM * ROOTS  (INCLUDING  *
2050  REM * COMPLEX ROOTS) OF A*
2060  REM * POLYNOMIAL WITH     *
2070  REM * REAL COEFFICIENTS. *
2080  REM *                     *
2090  REM * THE POLYNOMIAL IS  *
2100  REM * OF THE FORM:        *
2110  REM *                     *
2120  REM * X^N +               *
2130  REM * COF(0) * X^(N-1) + *
2140  REM * COF(1) * X^(N-2) + *
2150  REM *       .             *
2160  REM *       .             *
2170  REM * COF(N-2) * X +      *
2180  REM * COF(N-1) = 0        *
2190  REM *                     *
2200  REM *                     *
2210  REM * PARAMETERS:         *
2220  REM *                     *
2230  REM *   N - ORDER OF THE *
2240  REM *       POLYNOMIAL.   *
2250  REM *                     *
2260  REM * COF - AN ARRAY CON-*
2270  REM *       TAINING THE   *
2280  REM *       COEFFICIENTS *
2290  REM *       OF THE POLY- *
2300  REM *       NOMIAL.       *
2310  REM *       DIMENSIONED   *
2320  REM *       OF SIZE (N). *
2330  REM *                     *
2340  REM * A,B,                *
2350  REM * C,D - WORKING       *
2360  REM *       ARRAYS EACH   *
2370  REM *       DIMENSIONED   *
2380  REM *       OF SIZE (N). *
2390  REM *                     *
2400  REM * RR,                 *
2410  REM * RI - VECTORS CON-   *
2420  REM *      TAINING THE    *
2430  REM *      REAL AND       *
2440  REM *      IMAGINARY      *
2450  REM *      PARTS OF THE   *
2460  REM *      ROOTS FOUND.   *
2470  REM *      DIMENSIONED TO*
2480  REM *      SIZE (N).      *
2490  REM *                     *
2500  REM * PP,                 *
2510  REM * QQ - WORKING ARRAYS*
2520  REM *      DIMENSIONED TO*
2530  REM *      SIZE (N/2+1). *
2540  REM *                     *
2550  REM * THE SUBROUTINE      *
2560  REM * MAKES USE OF TWO    *
2570  REM * AUXILLARY ROUTINES *
2580  REM * INVOLVING NEWTON'S *
2590  REM * ITERATION METHOD.   *
2600  REM * IF THESE ROUTINES   *
2610  REM * FAIL TO CONVERGE    *
2620  REM * AFTER 100 ITERA-    *
2630  REM * TIONS, THE PROCESS *
2640  REM * IS TERMINATED.      *
2650  REM **********************
2660 :
2670 :
2680 M = N:P = 0:Q = 0
2690 L =  INT ((N + 1) / 2)
2700 :
2710  FOR I = 0 TO N - 1
2720 A(I) = COF(I)
2730  NEXT
2740 :
2750 :
2760  REM * FIRST FIND ROOTS    *
2770  REM * USING THE REDUCED   *
2780  REM * POLYNOMIAL.         *
2790 :
2800  FOR J = 0 TO L - 2
2810  GOSUB 3520
2820 PP(J) = P
2830 QQ(J) = Q
2840 M = M - 2
2850  IF M = 2 THEN  GOTO 3100
2860  FOR K = 0 TO M - 1
2870 A(K) = B(K)
2880  NEXT K
2890  IF M = 3 THEN  GOTO 2980
2900  NEXT J
2910 :
2920 :
2930  REM *THIS SECTION HANDLES*
2940  REM *THE LAST SINGLE ROOT*
2950  REM *FOR ODD ORDER        *
2960  REM *POLYNOMIALS.         *
2970 :
2980 X =  - B(0)
2990  GOSUB 3950
3000 PP(L - 2) = A(0) + X
3010 QQ(L - 2) =  - A(2) / X
3020  GOTO 3220
3030 :
3040 :
3050  REM *THIS SECTION HANDLES*
3060  REM *  THE LAST PAIR OF   *
3070  REM * ROOTS FOR AN EVEN   *
```

```
3080  REM * ORDER POLYNOMIAL.   *
3090 :
3100 PP(L - 1) = B(0)
3110 QQ(L - 1) = B(1)
3120 L = L + 1
3130 :
3140 :
3150  REM *THIS SECTION REFINES*
3160  REM * THE PREVIOUS ROOT   *
3170  REM * VALUES FOUND BY     *
3180  REM * DOING A FINAL       *
3190  REM * SET OF ITERATIONS   *
3200  REM * USING THE ORIGINAL  *
3210  REM * POLYNOMIAL.         *
3220 :
3230  FOR K = 0 TO N - 1
3240 A(K) = COF(K)
3250  NEXT K
3260  IF M = 4 THEN  GOTO 3310
3270 M = N
3280  GOSUB 3950
3290 RR(N - 1) = X
3300 RI(N - 1) = 0
3310 M = N
3320  FOR J = 0 TO L - 2
3330 P = PP(J)
3340 Q = QQ(J)
3350  GOSUB 3520
3360 D = P * P - 4 * Q
3370  IF D < 0 GOTO 3430
3380 RR(2 * J) = ( - P +  SQR (D)
     ) / 2.
3390 RR(2 * J + 1) = ( - P -  SQR
     (D)) / 2.
3400 RI(2 * J) = 0.
3410 RI(2 * J + 1) = 0.
3420  GOTO 3470
3430 RR(2 * J) =  - P / 2.
3440 RR(2 * J + 1) =  - P / 2.
3450 RI(2 * J) =  SQR ( - D) / 2.

3460 RI(2 * J + 1) =  -  SQR ( -
     D) / 2.
3470  NEXT J
3480 :
3490  RETURN
3500 :
3510 :
3520  REM *********************
3530  REM * THIS SUBROUTINE     *
3540  REM * FINDS A QUADRATIC   *
3550  REM * FACTOR OF  A        *
3560  REM * POLYNOMIAL OF ORDER*
3570  REM * 'M'.                *
3580  REM *********************
3590 :
3600 B(0) = A(0) - P
3610 B(1) = A(1) - P * B(0) - Q
3620 C(0) =  - 1.
3630 C(1) =  - B(0) + P
3640 D(0) = 0.
3650 D(1) =  - 1.
3660  FOR I = 2 TO M - 3
3670 B(I) = A(I) - P * B(I - 1) -
     Q * B(I - 2)
3680 C(I) =  - B(I - 1) - P * C(I
      - 1) - Q * C(I - 2)
3690 D(I) = C(I - 1)
3700  NEXT I
3710 R = A(M - 2) - P * B(M - 3) -
     Q * B(M - 4)
3720 S =  + A(M - 1) - Q * B(M -
     3)
3730 RQ =  - B(M - 3) - P * C(M -
     3) - Q * C(M - 4)
3740 RP =  - B(M - 3) - P * C(M -
     3) - Q * C(M - 4)
3750 RQ =  - B(M - 4) - P * D(M -
     3) - Q * D(M - 4)
3760 SP =  - Q * C(M - 3)
3770 SQ =  - B(M - 3) - Q * D(M -
     3)
3780 DE = RP * SQ - SP * RQ
3790 DP = (S * RQ - R * SQ) / DE
3800 DQ = (R * SP - S * RP) / DE
3810  IF  ABS (DP) > 0.001 *  ABS
     (P) GOTO 3860
3820  IF  ABS (DQ) > 0.001 *  ABS
     (Q) GOTO 3860
3830 P = P + DP
3840 Q = Q + DQ
3850  RETURN
3860 ICT = ICT + 1
3870  IF ICT > 100 THEN  GOTO 391
     0
3880 P = P + DP
3890 Q = Q + DQ
3900  GOTO 3600
3910  PRINT "NO SOLUTION AFTER 10
     0 ITERATIONS"
3920  RETURN
3930 :
3940 :
3950  REM *********************
3960  REM * THIS SUBROUTINE IS *
3970  REM * USED TO EXTRACT A  *
3980  REM * SINGLE ROOT OF THE *
3990  REM * POLYNOMIAL BY THE  *
4000  REM * NEWTON ITERATION   *
4010  REM * METHOD.            *
4020  REM *********************
4030 :
4040 ICT = 0
4050 :
4060 F = X ^ M + A(M - 1)
4070 DF = M * X ^ (M - 1)
```

```
4080  FOR I = 0 TO M - 2
4090 F = F + A(I) * X ^ (M - 1 -
     I)
4100 DF = DF + A(I) * (M - 1 - I)
      * X ^ (M - 2 - I)
4110  NEXT
4120 :
4130 DX =  - F / DF
4140  IF  ABS (DX) > 0.0001 *  ABS
     (X) THEN  GOTO 4170
4150 X = X + DX
4160  RETURN
4170 ICT = ICT + 1
4180  IF ICT > 100 GOTO 4210
4190 X = X + DX
4200  GOTO 4060
4210  PRINT "OVER 100 ITERATIONS"
4220  RETURN
```

The output from this program is as follows:

```
THE ROOTS ARE AS FOLLOWS
---------------------------
  REAL              IMAGINARY
  PART                 PART
---------------------------
3                 0
-4                0
1                 3
1                 -3
-5                0
---------------------------
```

This program required 8.5 seconds to run on an Apple II computer.

2.10 CONSIDERATIONS IN THE SELECTION OF AN ALGORITHM FOR THE SMALL COMPUTER

Although it is impossible to state universal rules that will guide the user in selecting the best method for finding the root for a particular equation, a few basic guidelines do exist. These are as follows:

1. **Consider the nature of the problem and its roots.** If the problem to be solved is transcendental, the number of possible solutions will, in general, not be known. Also, the approximate location for roots will be unknown, making solution by Newton's method somewhat difficult. For this reason, the binary search method coupled with a method like that of false position may be highly desirable. If, on the other hand, something is known about the nature of the root or roots, the method of Newton iteration will usually provide a faster convergence to a usable solution. It should be kept in mind that Newton's method has difficulty handling situations in which the slope is zero or in which the function is discontinuous. Under these conditions, other methods perform better. Frequently, a rough sketch of the transcendental function may provide considerable

insight into a suitable starting value for an iterative method. If the function is not easily differentiable, the positive advantages of Newton's method can still be exploited by using a secant approximation to the derivative.

If the problem to be solved is a polynomial, the user has far more information to use in selecting a solution algorithm. For example, the user will know how many roots there will be, and can frequently tell how many will be positive real and how many will be negative real. This information is often useful in selecting an initial guess for an iterative method. One method that can be used to find an estimate of the smallest root of a polynomial is to truncate the higher-degree terms so that only a quadratic term remains. The smallest root of the quadratic is a good estimate for the smallest root of the polynomial. A similar method can be used to find an estimate for the largest root of the polynomial. If the polynomial is truncated by dropping all lower-degree terms, then the largest root of the surviving quadratic term will be an estimate for the largest root of the polynomial. If it is suspected that some of the roots may be complex, it is best to use methods like Lin's method or Bairstow's method, since they will yield complex roots without the need for complex algebraic manipulation.

2. **Consider the accuracy required.** The higher-order methods, such as Newton's method, will usually provide a higher level of accuracy than the binary search method or the method of false position. Yet there are times when a tradeoff between accuracy and overall solution utility more than justifies the selection of a lower-order method. It should also be kept in mind that methods using a reduced polynomial to find more than one root may provide results that are less and less accurate as further levels of reduction are used. For this reason, methods like Lin's method and Bairstow's method should always have a final iteration cycle that improves a root value using the full, original polynomial. An illustration of this is presented in Example 2-2.
3. **Consider the time involved.** In selecting an algorithm, it is well to keep in mind that an upper bound on the number of iterations required to reduce the interval of uncertainty by binary search is easily predicted. On the other hand, with Newton's method the number of iterations required is not predictable. Yet, when it converges, Newton's method is often quite rapid. For problems in which the time for calculation of the function value and its derivatives is extensive, the method of Newton iteration is clearly preferred.
4. **Consider the computer space required.** The Bairstow method presented in Example 2-2 uses far more computer storage space in find-

ing the roots of a polynomial of order N than would be required for a method like Newton's method. For this program the array space required is about $8*N$. Yet this method does find all roots whether complex or pure real and is thus much preferred for most general applications if computer storage space is available.

REFERENCES

1. GROVE, WENDELL E., *Brief Numerical Methods*, Prentice-Hall, Inc., Englewood Cliffs, N.J., 1966.
2. LA FARA, ROBERT L., *Computer Methods for Science and Engineering*, Hayden Book Company, Inc., Rochelle Park, N.J., 1973.
3. McCALLA, THOMAS R., *Introduction to Numerical Methods and FORTRAN Programming*, John Wiley & Sons, Inc., New York, 1967.
4. PALL, GABRIEL A., *Introduction to Scientific Computing*, Appleton-Century-Crofts, Educational Division, Meredith Corp., New York, 1971.
5. RALSTON, ANTHONY, *A First Course in Numerical Analysis*, McGraw-Hill Book Co., New York, 1965.
6. RALSTON, ANTHONY, and H.S. WILF, *Mathematical Methods for Digital Computers*, John Wiley & Sons, Inc., New York, 1967.
7. SALVADORI, MARIO G., and MELVIN L. BARON, *Numerical Methods in Engineering*, Prentice-Hall, Inc., Englewood Cliffs, N.J., 1961.
8. TURNER, L.R., "Solution of Nonlinear Systems," *Annals of the New York Academy of Sciences*, Vol. 86, 1960, pp. 817-827.
9. WILLIAMS, P.W., *Numerical Computation*, Harper & Row Publishers, Inc., New York, 1972.

The 8086 microprocessor. (*Photo courtesy of Intel Corporation.*)

Roots of simultaneous equations 3

The solution of linear simultaneous equations is, perhaps, one of the most common algebraic problems encountered in engineering calculations. The general formulation for this problem will be

$$a_{11}x_1 + a_{12}x_2 + \cdots + a_{1n}x_n = c_1$$

$$a_{21}x_1 + a_{22}x_2 + \cdots + a_{2n}x_n = c_2$$

$$\cdots$$

$$a_{n1}x_1 + a_{n2}x_2 + \cdots + a_{nn}x_n = c_n$$

In this formulation, the n equations must be linearly independent in order for a unique solution to exist. The necessary and sufficient condition for this to occur is that the determinant of the coefficient matrix must not be zero. Solution algorithms for this type of problem may be either direct or indirect. The fact that direct methods can, for this problem situation, be systematized makes them quite popular. In this chapter the direct methods will be considered first, and the indirect (iterative) methods will be considered last.

3.1 GAUSSIAN ELIMINATION METHOD

The most frequently used direct methods are based on a procedure for reducing the equation system into a "triangular" form so that one of the equations contains only one of the unknowns, and each of the next equations contains only one additional new unknown. In hand-calculation techniques, triangularization is achieved by addition and subtraction of various equations after they have been multiplied by appropriate constant factors. Although this procedure may be somewhat haphazard when implemented by hand, a systematic scheme for computer implementation can be established. The method of Gaussian elimination is one such method. This procedure starts by normalizing the first equation by dividing each of its coefficients by a_{11}. Next, this first equation is multiplied by the leading coefficient $a_{i,1}$ of each of the other equations and is subtracted from each successive equation. The result will be the elimination of the first variable from all equations except the first. Next, using the last $n - 1$ equations and the same procedure, the second variable is eliminated from the last $n - 2$ equations. The procedure is repeated until after n stages the triangular form is complete. Mathematically, this procedure can be stated as follows:

At the kth state in the elimination process the new, normalized coefficients for the kth equation are

$$b_{k,j} = \frac{a_{k,j}}{a_{kk}}$$

and the new coefficients in the equations that follow will be:

$$b_{i,j} = a_{i,j} - a_{i,k} b_{k,j} \qquad i > k$$

In performing this process it should be kept in mind that the $a_{i,j}$ coefficients of the lower equations change during each stage of the process. Thus the $b_{i,j}$ coefficients for one stage become the $a_{i,j}$ coefficients that are used for the next stage. This procedure is best illustrated by a simple illustration.

Suppose it is desired to solve the following equations by Gaussian elimination:

$$x_1 + x_2 + x_3 - x_4 = 2$$

$$x_1 - x_2 - x_3 + x_4 = 0$$

$$2x_1 + x_2 - x_3 + 2x_4 = 9$$

$$3x_1 + x_2 + 2x_3 - x_4 = 7$$

For ease of manipulation, the rows in this system will be identified by letter and only the coefficient array will be written. The original array to be manipulated is thus

Row	Array				
A_1	1	1	1	-1	2
A_2	1	-1	-1	2	0
A_3	2	1	-1	2	9
A_4	3	1	2	-1	7

After the elimination of x_1 terms, the array will be

Row	Array				
$B_1 = A_1/1$	1	1	1	-1	2
$B_2 = A_2 - B_1$	0	-2	-2	2	-2
$B_3 = A_3 - 2B_1$	0	-1	-3	4	5
$B_4 = A_4 - 3B_1$	0	-2	-1	2	1

After the elimination of x_2 terms, the array will be

Row	Array				
B_1	1	1	1	-1	2
$C_2 = B_2/-2$	0	1	1	-1	1
$C_3 = B_3 + C_2$	0	0	-2	3	6
$C_4 = B_4 + 2C_2$	0	0	1	0	3

After the elimination of x_3 terms, the array will be

Row	Array				
B_1	1	1	1	-1	2
C_2	0	1	1	-1	1
$D_3 = C_3/(-2)$	0	0	1	$-\frac{3}{2}$	-3
$D_4 = C_4 - D_3$	0	0	0	$\frac{3}{2}$	6

After reducing the coefficients on the last row, the array will be

Row	Array				
B_1	1	1	1	-1	2
C_2	0	1	1	-1	1
D_3	0	0	1	$-\frac{3}{2}$	-3
$E_4 = D_4/(\frac{3}{2})$	0	0	0	1	4

The array can now be written in equation form as

$$x_1 + x_2 + x_3 - x_4 = 2$$
$$x_2 + x_3 - x_4 = 1$$
$$x_3 - \tfrac{3}{2} x_4 = -3$$
$$x_4 = 4$$

By applying the process of back substitution, the results

$$x_1 = 1$$
$$x_2 = 2$$
$$x_3 = 3$$
$$x_4 = 4$$

can be found.

This example clearly illustrates the fact that it is desirable to reduce all off-diagonal elements to zero. This procedure is called *diagonalization* and is an improvement over traingularization.

3.2 GAUSS–JORDAN ELIMINATION METHOD

The Gauss–Jordan elimination method provides a systematic means for the diagonalization of a system of linear simultaneous equations. The only mathematical difference between this direct method and the previous direct method is that $i \neq k$ is substituted for $i > k$. The kth row is called the *pivot* row. In the previous method, only the equations below the pivot row were manipulated, whereas in the Gauss–Jordan method the manipulation takes place both above and below the pivot row. To illustrate this procedure, the previous example will be worked using the Gauss–Jordan procedure. The original array to be manipulated is

Row	Array				
A_1	1	1	1	-1	2
A_2	1	-1	-1	1	0
A_3	2	1	-1	2	9
A_4	3	1	2	-1	7

After the elimination of the x_1 terms, the array will be

Row	Array				
$B_1 = A_1/1$	1	1	1	-1	2
$B_2 = A_2 - B_1$	0	-2	-2	2	-2
$B_3 = A_3 - 2B_1$	0	-1	-3	4	5
$B_4 = A_4 - 3B_1$	0	-2	-1	2	1

Up to this point the procedure is identical to that for the Gaussian elimination technique. After the elimination of the x_2 terms, the array will be

Row	Array				
$C_1 = B_1 - C_2$	1	0	0	0	1
$C_2 = B_2/(-2)$	0	1	1	-1	1
$C_3 = B_3 + C_2$	0	0	-2	3	6
$C_4 = B_4 + 2C_2$	0	0	1	0	3

This new array is, of course, different from that found for the third stage of the Gaussian elimination technique. After the elimination of the x_3 terms, the array will be

Row	Array				
$D_1 = C_1 - (0)D_3$	1	0	0	0	1
$D_2 = C_2 - D_3$	0	1	0	$\frac{1}{2}$	4
$D_3 = C_3/(-2)$	0	0	1	$-\frac{3}{2}$	-3
$D_4 = C_4 - D_3$	0	0	0	$\frac{3}{2}$	6

Finally, after the elimination of the x_4 terms in all rows except the last, the array will be

Row	Array				
$E_1 = D_1 + (0)E_4$	1	0	0	0	1
$E_2 = D_2 - \frac{1}{2}E_4$	0	1	0	0	2
$E_3 = D_3 + \frac{3}{2}E_4$	0	0	1	0	3
$E_4 = D_4/(\frac{3}{2})$	0	0	0	1	4

Clearly, for this method the answers are easier to extract. The disadvantage to this method is its need for additional calculations.

A potential pitfall in the two foregoing methods can occur if any pivot element is zero. When this happens, the normalization of the pivot row cannot be accomplished. Since it is possible to change the order of the equations in the system, this device can provide a way to circumvent the zero pivot element problem. Indeed, it can be shown that the greatest overall computational accuracy is achieved when the pivot element has the greatest magnitude. Thus the row with a zero or small pivot element should be exchanged with the row below it that has the largest element in the same column. When this is done, the process is known as *partial pivoting.* Such a process is especially important when handling large numbers of simultaneous equations for reasons of improved accuracy. *Complete pivoting* is the process of interchanging columns as well as rows.

EXAMPLE 3-1 Suppose that it is desired to solve the following system of four linear equations:

$$-2.0x_1 + 1.1x_2 - 2.0x_3 - 1.8x_4 = 1.0$$

$$3.2x_1 + 2.1x_2 + 3.2x_3 + 2.2x_4 = 1.0$$

$$3.4x_1 + 2.3x_2 + 4.1x_3 + 3.2x_4 = 6.0$$

$$2.6x_1 + 1.1x_2 - 3.2x_3 + 2.4x_4 = -7.0$$

The method to be used will be that of Gauss–Jordan elimination with partial pivoting. A BASIC program that implements this problem is given next. This program is constructed in two parts. In the first part the array *A* consisting of 4 rows and 5 columns is loaded with the coefficient matrix and the right sides of the preceding equation system. It also establishes a working array *B* of size equivalent to *A*. The subroutine starting at line 2000 implements the actual Gauss–Jordan elimination process with partial pivoting. As a result of the elimination process, the array *A* contains the identity matrix in its first n columns and the solution values in its $n + 1$ column.

```
1000  REM *********************
1010  REM *THIS PROGRAM FINDS  *
1020  REM *THE SOLUTION TO A   *
1030  REM *SET OF SIMULTANEOUS *
1040  REM *LINEAR ALGEBRAIC    *
1050  REM *EQUATIONS BY THE    *
1060  REM *GAUSS-JORDAN METHOD *
1070  REM *USING PARTIAL       *
1080  REM *PIVOTING.           *
1090  REM *********************
1100 :
1110 :
1120  REM **SET UP THE ARRAY**
1130 :
1140  DIM A(5,6),B(5,6)
1150 NROW = 4:NCOL = 5
1160  FOR J = 1 TO NROW
1170  FOR I = 1 TO NCOL
1180  READ A(J,I)
1190  NEXT I
1200  NEXT J
1210  DATA  -2.0,1.1,-2.0,-1.8,1
1220  DATA  3.2,2.1,3.2,2.2,1
1230  DATA  3.4,2.3,4.1,3.2,6
1240  DATA  2.6,1.1,-3.2,2.4,-7
1250 :
1260 :
1270  REM **APPLY GAUSS-JORDAN**
1280  REM **ON RETURN THE ANSWERS

1290  REM **WILL BE FOUND IN THE
1300  REM **LAST COLUMN OF THE
1310  REM **MATRIX "A."
1320  GOSUB 2000
1330 :
1340 :
1350  REM **WRITE THE ANSWERS**
1360 :
1370  PRINT "------------------"
1380  PRINT "ANSWERS ARE"
1390  PRINT "------------------"
1400  FOR J = 1 TO NROW
1410  PRINT "X(";J;")=";A(J,NCOL)

1420  NEXT J
1430  PRINT "------------------"
1440  PRINT
1450 :
1460  END
1470 :
1480 :
2000  REM *********************
2010  REM * THIS SUBROUTINE    *
2020  REM * APPLIES THE METHOD *
2030  REM * OF GAUSS-JORDAN    *
2040  REM * ELIMINATION TO A   *
2050  REM * MATRIX USING       *
2060  REM * PARTIAL PIVOTING.  *
2070  REM *                    *
2080  REM *   PARAMETERS:      *
2090  REM *                    *
2100  REM *    A - THE ORIGINAL*
2110  REM *        AUGMENTED   *
2120  REM *        MATRIX      *
2130  REM *        DIMENSIONED *
2140  REM *        (NROW+1,    *
2150  REM *           NCOL+1). *
2160  REM *                    *
2170  REM *    B - A WORKING   *
2180  REM *        MATRIX OF   *
2190  REM *        SIZE SIMILAR*
2200  REM *        TO "A."     *
2210  REM *                    *
2220  REM *  NROW- THE NUMBER   *
2230  REM *        OF ROWS IN  *
2240  REM *        THE A MATRIX*
2250  REM *                    *
2260  REM *  NCOL- THE NUMBER   *
2270  REM *        OF COLUMNS  *
2280  REM *        IN THE "A"  *
2290  REM *        MATRIX.     *
2300  REM *                    *
2310  REM *********************
2320 :
```

```
2330 :
2340  REM **DO THE PROBLEM   **
2350  REM **IN STAGES        **
2360 :
2370  FOR K = 1 TO NROW
2380 :
2390 :
2400  REM **FIND LARGEST PIVOT**
2410 :
2420 PIVOT = A(K,K):IL = K
2430  FOR L = K + 1 TO NROW
2440  IF  ABS (A(L,K)) <  ABS (PI
     VOT) THEN  GOTO 2470
2450 PIVOT = A(L,K)
2460 IL = L
2470  NEXT L
2480 :
2490 :
2500  REM **TRADE ROWS TO GET**
2510  REM **LARGEST PIVOT    **
2520 :
2530  FOR LL = 1 TO NCOL
2540 TEMP = A(K,LL)
2550 A(K,LL) = A(IL,LL)
2560 A(IL,LL) = TEMP
2570  NEXT LL
2580 :
2590 :
2600  REM **NORMALIZE PIVOT ROW*
2610 :
2620  FOR J = 1 TO NCOL
2630 B(K,J) = A(K,J) / PIVOT
2640  NEXT J
2650 :
2660  REM **DO GAUSS-JORDAN**
2670  REM * ELIMINATION STEP*
2680 :
2690  FOR I = 1 TO NROW
2700  IF I = K GOTO 2740
2710  FOR J = 1 TO NCOL
2720 B(I,J) = A(I,J) - A(I,K) * B
     (K,J)
2730  NEXT J
2740  NEXT I
2750 :
2760 :
2770  REM **UPDATE 'A' MATRIX**
2780  REM **WITH 'B' MATRIX  **
2790 :
2800  FOR I = 1 TO NROW
2810  FOR J = 1 TO NCOL
2820 A(I,J) = B(I,J)
2830  NEXT J
2840  NEXT I
2850  NEXT K
2860  RETURN
```

The output from this program is as follows:

```
------------------
ANSWERS ARE
------------------
X(1)=-5.20995646
X(2)=1.42625267
X(3)=1.64909869
X(4)=4.27255194
------------------
```

This program required less than 4 seconds to run on an Apple II computer.

3.3 FINDING A MATRIX INVERSE BY THE GAUSS-JORDAN ELIMINATION METHOD

The Gauss–Jordan elimination method provides a systematic means for transforming a matrix into the identity matrix. This process is exactly what would be required to find the inverse of the original matrix. To implement this process, the user should augment the original matrix with the identity matrix. Then at the end of the elimination process when the original matrix has been transformed into the identity matrix, the augmented part of the matrix that started as the identity will become the inverse of the original matrix. The previous text example will be worked using the Gauss–Jordan procedure to illustrate this inverse process. The original matrix is first written along with the augmented identity matrix as

Row	Array							
A_1	1	1	1	-1	1	0	0	0
A_2	1	-1	-1	1	0	1	0	0
A_3	2	1	-1	2	0	0	1	0
A_4	3	1	2	-1	0	0	0	1

After the elimination of the x_1 terms, the array will be

Row	Array							
$B_1 = A_1/1$	1	1	1	-1	1	0	0	0
$B_2 = A_2 - B_1$	0	-2	-2	2	-1	1	0	0
$B_3 = A_3 - 2B_1$	0	-1	-3	4	-2	0	1	0
$B_4 = A_4 - 3B_1$	0	0	-1	2	-3	0	0	1

After the elimination of the x_2 terms, the array will be

Row	Array							
$C_1 = B_1 - C_2$	1	0	0	0	$\frac{1}{2}$	$\frac{1}{2}$	0	0
$C_2 = B_2/(-2)$	0	1	1	-1	$\frac{1}{2}$	$-\frac{1}{2}$	0	0
$C_3 = B_3 + C_2$	0	0	-2	3	$-\frac{3}{2}$	$-\frac{1}{2}$	1	0
$C_4 = B_4 + 2C_2$	0	0	1	0	-2	-1	0	1

After the elimination of the x_3 terms, the array will be

Row	Array							
$D_1 = C_1 - (0)D_3$	1	0	0	0	$\frac{1}{2}$	$\frac{1}{2}$	0	0
$D_2 = C_2 - D_3$	0	1	0	$\frac{1}{2}$	$-\frac{1}{4}$	$-\frac{3}{4}$	$\frac{1}{2}$	0
$D_3 = C_3/(-2)$	0	0	1	$-\frac{3}{2}$	$\frac{3}{4}$	$\frac{1}{4}$	$-\frac{1}{2}$	0
$D_4 = C_4 - D_3$	0	0	0	$\frac{3}{2}$	$-\frac{11}{4}$	$-\frac{5}{4}$	$\frac{1}{2}$	1

Finally, after the elimination of the x_4 terms in all rows except the last, the array will be

Row	Array							
$E_1 = D_1 + (0)\,E_4$	1	0	0	0	$\frac{1}{2}$	$\frac{1}{2}$	0	0
$E_2 = D_2 - \frac{1}{2}\,E_4$	0	1	0	0	$\frac{2}{3}$	$-\frac{1}{3}$	$\frac{1}{3}$	$-\frac{1}{3}$
$E_3 = D_3 + \frac{3}{2}\,E_4$	0	0	1	0	-2	-1	0	1
$E_4 = D_4/(\frac{3}{2})$	0	0	0	1	$-\frac{11}{6}$	$-\frac{5}{6}$	$\frac{1}{3}$	$\frac{2}{3}$

The right portion of this array now contains the inverse of the original matrix. This can easily be verified by multiplying this matrix times the original matrix to see if the identity matrix results. Thus,

$$\begin{bmatrix} \frac{1}{2} & \frac{1}{2} & 0 & 0 \\ \frac{2}{3} & -\frac{1}{3} & \frac{1}{3} & -\frac{1}{3} \\ -2 & -1 & 0 & 0 \\ -\frac{11}{6} & -\frac{5}{6} & \frac{1}{3} & \frac{2}{3} \end{bmatrix} \begin{bmatrix} 1 & 1 & 1 & -1 \\ 1 & -1 & -1 & 1 \\ 2 & 1 & -1 & 2 \\ 3 & 1 & 2 & -1 \end{bmatrix} = \begin{bmatrix} 1 & 0 & 0 & 0 \\ 0 & 1 & 0 & 0 \\ 0 & 0 & 1 & 0 \\ 0 & 0 & 0 & 1 \end{bmatrix}$$

As was the case for the solution of simultaneous equations, the accuracy of this process can be maximized through the use of pivoting.

EXAMPLE 3-2 Suppose that it is desired to find the inverse of the coefficient matrix from Example 3-1:

$$\begin{bmatrix} -2.0 & 1.1 & -2.0 & -1.8 \\ 3.2 & 2.1 & 3.2 & 2.2 \\ 3.4 & 2.3 & 4.1 & 3.2 \\ 2.6 & 1.1 & -3.2 & 2.4 \end{bmatrix}$$

The method to be used will be that of Gauss–Jordan elimination with partial pivoting. A BASIC program that implements this problem is given next. This program makes use of the same Gauss–Jordan subroutine used in Example 3-1. The array A consisting of 4 rows and 8 columns is loaded with the coefficient matrix and is augmented by the identity matrix on the right side.

```
1000  REM *********************
1010  REM * THIS PROGRAM FINDS *
1020  REM * THE INVERSE OF A   *
1030  REM * MATRIX BY THE      *
1040  REM * GAUSS-JORDAN METHOD*
1050  REM * USING PARTIAL      *
1060  REM * PIVOTING.          *
1070  REM *********************
1080 :
1090 :
1100  DIM A(5,9),B(5,9),G$(5)
1110 NROW = 4:NCOL = 8
1120 :
1130  REM **SET UP THE ARRAY**
1140  FOR J = 1 TO NROW
1150  FOR I = 1 TO NCOL
1160  READ A(J,I)
1170  NEXT I
1180  NEXT J
1190  DATA  -2,1.1,-2,-1.8,1,0,0,
     0
1200  DATA  3.2,2.1,3.2,2.2,0,1,0
     ,0
1210  DATA  3.4,2.3,4.1,3.2,0,0,1
     ,0
1220  DATA  2.6,1.1,-3.2,2.4,0,0,
     0,1
1230 :
1240 :
1250  REM **APPLY GAUSS-JORDAN**
1260  GOSUB 2000
1270  REM **ON RETURN THE
1280  REM **INVERSE MATRIX
1290  REM **WILL BE IN THE
1300  REM **AUGMENTED,
1310  REM **RIGHT-HAND SIDE
1320  REM **OF MATRIX "A."
1330 :
1340 :
1350  REM **WRITE THE ANSWERS**
1360 :
1370  PRINT "-------------------
     -------------------"
1380  PRINT "THE INVERSE MATRIX I
     S"
1390  PRINT "-------------------
     -------------------"
1400  GOSUB 4000
1410  PRINT "-------------------
     -------------------"
1420  PRINT
1430 :
1440  END
1450 :
1460 :
2000  REM *********************
2010  REM * THIS SUBROUTINE    *
2020  REM * APPLIES THE METHOD *
2030  REM * OF GAUSS-JORDAN    *
2040  REM * ELIMINATION TO A   *
2050  REM * MATRIX USING       *
2060  REM * PARTIAL PIVOTING.  *
2070  REM *                    *
2080  REM *   PARAMETERS:      *
2090  REM *                    *
2100  REM *    A - THE ORIGINAL*
2110  REM *        AUGMENTED   *
2120  REM *        MATRIX      *
2130  REM *        DIMENSIONED *
2140  REM *        (NROW+1,    *
2150  REM *           NCOL+1). *
2160  REM *                    *
2170  REM *    B - A WORKING   *
2180  REM *        MATRIX OF   *
2190  REM *        SIZE SIMILAR*
2200  REM *        TO "A."     *
2210  REM *                    *
2220  REM *  NROW- THE NUMBER   *
2230  REM *        OF ROWS IN  *
2240  REM *        THE A MATRIX*
2250  REM *                    *
2260  REM *  NCOL- THE NUMBER   *
2270  REM *        OF COLUMNS  *
2280  REM *        IN THE "A"  *
2290  REM *        MATRIX.     *
2300  REM *                    *
2310  REM *********************
2320 :
2330 :
2340  REM **DO THE PROBLEM   **
2350  REM **IN STAGES        **
2360 :
2370  FOR K = 1 TO NROW
2380 :
2390 :
2400  REM **FIND LARGEST PIVOT**
2410 :
2420 PIVOT = A(K,K):IL = K
2430  FOR L = K + 1 TO NROW
2440  IF  ABS (A(L,K)) <  ABS (PI
     VOT) THEN  GOTO 2470
2450 PIVOT = A(L,K)
2460 IL = L
2470  NEXT L
2480 :
2490 :
2500  REM **TRADE ROWS TO GET**
2510  REM **LARGEST PIVOT    **
2520 :
2530  FOR LL = 1 TO NCOL
2540 TEMP = A(K,LL)
2550 A(K,LL) = A(IL,LL)
2560 A(IL,LL) = TEMP
2570  NEXT LL
2580 :
2590 :
```

```
2600  REM **NORMALIZE PIVOT ROW*
2610 :
2620  FOR J = 1 TO NCOL
2630 B(K,J) = A(K,J) / PIVOT
2640  NEXT J
2650 :
2660  REM **DO GAUSS-JORDAN**
2670  REM * ELIMINATION STEP*
2680 :
2690  FOR I = 1 TO NROW
2700  IF I = K GOTO 2740
2710  FOR J = 1 TO NCOL
2720 B(I,J) = A(I,J) - A(I,K) * B
     (K,J)
2730  NEXT J
2740  NEXT I
2750 :
2760 :
2770  REM **UPDATE 'A' MATRIX**
2780  REM **WITH 'B' MATRIX  **
2790 :
2800  FOR I = 1 TO NROW
2810  FOR J = 1 TO NCOL
2820 A(I,J) = B(I,J)
2830  NEXT J
2840  NEXT I
2850  NEXT K
2860  RETURN
2870 :
2880 :
4000  REM *********************
4010  REM * THIS SUBROUTINE    *
4020  REM * PRINTS THE RESULTS *
4030  REM *********************
4040 :
4050 :
4060  FOR J = 1 TO 4
4070  FOR I = 1 TO 4
4080 F(I) = A(J,I + 4)
4090  NEXT I
4100  GOSUB 4170
4110  PRINT G$(1); TAB( 11);G$(2)
     ; TAB( 21);G$(3); TAB( 31);G
     $(4)
4120  NEXT J
4130 :
4140  RETURN
4150 :
4160 :
4170  REM ********************
4180  REM * THIS SUBROUTINE   *
4190  REM * SORTS SCIENTIFIC  *
4200  REM * NOTATION FOR PRINT*
4210  REM * CLARITY.          *
4220  REM ********************
4230 :
4240 :
4250  FOR I = 1 TO 4
4260  IF  ABS (F(I)) > 0.01 THEN
      GOTO 4290
4270 G$(I) =  LEFT$ ( STR$ (F(I))
     ,2) +  RIGHT$ ( STR$ (F(I)),
     4)
4280  GOTO 4310
4290 G$(I) =  STR$ (F(I))
4300 G$(I) =  LEFT$ (G$(I),8)
4310  NEXT I
4320  RETURN
```

The output from this program is as follows:

```
---------------------------------------
THE INVERSE MATRIX IS
---------------------------------------
-.230436  1.132705  -.951535  .0575733
.4148525  .0407691  .1882829  .0227238
-.043156  4.E-03    .0931866  -.161167
1.E-03    -1.23916  1.068782  .1289909
---------------------------------------
```

This program required less than 6 seconds to run on an Apple II computer.

3.4 CHOLESKY'S METHOD FOR SIMULTANEOUS LINEAR EQUATIONS

One disadvantage of the Gauss–Jordan method for simultaneous linear equations is that the working storage space required for its implementation is roughly twice that required to store the augmented coefficient matrix. Cholesky's method, also known as Crout's method, can be implemented in such a way that this disadvantage is overcome. This method is also more economical of computer time than is the Gauss–Jordan method and thus has much promise for use on the small computer.

The basis for Cholesky's method is that the augmented matrix

$$A = \begin{bmatrix} a_{11} & a_{12} & a_{13} & a_{14} & a_{15} \\ a_{21} & a_{22} & a_{23} & a_{24} & a_{25} \\ a_{31} & a_{32} & a_{33} & a_{34} & a_{35} \\ a_{41} & a_{42} & a_{43} & a_{44} & a_{45} \end{bmatrix}$$

representing the system of equations can be reduced to an equivalent upper triangular system of the form

$$U = \begin{bmatrix} 1 & u_{12} & u_{13} & u_{14} & u_{15} \\ 0 & 1 & u_{23} & u_{24} & u_{25} \\ 0 & 0 & 1 & u_{34} & u_{35} \\ 0 & 0 & 0 & 1 & u_{45} \end{bmatrix}$$

If this form can be found, the solution to the equation system can easily be found by back substitution starting with the last equation. One way to achieve this upper triangular form is to use Gaussian elimination. Another way is to find a transformation matrix that, when premultiplied times the U matrix, transforms it into the original A matrix. It can be shown that the transformation matrix will be a special lower triangular matrix of the form

$$L = \begin{bmatrix} l_{11} & 0 & 0 & 0 \\ l_{21} & l_{22} & 0 & 0 \\ l_{31} & l_{32} & l_{33} & 0 \\ l_{41} & l_{42} & l_{43} & l_{44} \end{bmatrix}$$

Thus the result will be

$$[L]\,[U] = [A]$$

or

$$\begin{bmatrix} l_{11} & 0 & 0 & 0 \\ l_{21} & l_{22} & 0 & 0 \\ l_{31} & l_{32} & l_{33} & 0 \\ l_{41} & l_{42} & l_{43} & l_{44} \end{bmatrix} \begin{bmatrix} 1 & u_{12} & u_{13} & u_{14} & u_{15} \\ 0 & 1 & u_{23} & u_{24} & u_{25} \\ 0 & 0 & 1 & u_{34} & u_{35} \\ 0 & 0 & 0 & 1 & u_{45} \end{bmatrix}$$

$$= \begin{bmatrix} a_{11} & a_{12} & a_{13} & a_{14} & a_{15} \\ a_{21} & a_{22} & a_{23} & a_{24} & a_{25} \\ a_{31} & a_{32} & a_{33} & a_{34} & a_{35} \\ a_{41} & a_{42} & a_{43} & a_{44} & a_{45} \end{bmatrix}$$

By performing the multiplication of these matrices and comparing the terms on each side of the equality, the relationships for determining the coefficients of L and U can be found. For example,

$$l_{i1} = a_{i1}, \qquad \text{for} \quad i = 1, 2, \ldots, n$$

gives the first column of L. Once this is known, the first row of U may be found by

$$u_{1j} = \frac{a_{1j}}{l_{11}}, \qquad \text{for} \quad j = 2, 3, \ldots, n + 1$$

Next the second column of L and then the second row of U can be found as

$$l_{i2} = a_{i2} - l_{21} u_{22}, \qquad \text{for} \quad i = 2, 3, \ldots, n$$

and

$$u_{2j} = \frac{a_{2j} - l_{21} u_{1j}}{l_{22}}, \qquad \text{for} \quad j = 3, 4, \ldots, n + 1$$

Next the third column of L and the third row of U are found, then the fourth column of L and the fourth row of U, and so on. The process continues until all rows and columns are found. Since each step uses only information previously found, the storage space required can be streamlined somewhat by overlaying both U and L onto the A matrix. This may be done as long as the zeros and ones of L and U are not carried along. For this situation the first column of L is automatically set without an assignment. The remaining assignment relationships are as follows:

1. For the first row, $a_{1j} = a_{1j}/a_{11}$, for $j = 2, 3, \ldots, n+1$.
2. For the mth column and mth row (for $m = 2, 3, \ldots, n$):

   (a) For the l_{im} columns in turn,

$$a_{im} = a_{im} - \sum_{k=1}^{m-1} a_{ik}\, a_{km}, \qquad \text{for} \quad i = m, \ldots, n$$

   (b) For the u_{mj} rows in turn,

$$a_{mj} = \frac{a_{mj} - \sum_{k=1}^{m-1} a_{mk}\, a_{kj}}{a_{mm}}, \qquad \text{for} \quad j = m+1, \ldots, n+1$$

Once the U matrix is known, the x_i solution values can be found by back substitution using

$$x_n = a_{n,n+1}$$

$$x_i = a_{i,n+1} - \sum_{k=i+1}^{n} a_{ik}\, x_k, \qquad \text{for} \quad i = n-1, n-2, \ldots, 1$$

Since this procedure contains a division by a_{mm} in the row calculation, it is important that these terms not equal zero. Indeed, it can be shown that the best accuracy is ensured when the a_{mm} values are as large as possible. This suggests that partial pivoting similar to that used in the Gauss–Jordan method is useful.

If the A matrix is symmetrical, the number of calculations and the amount of storage space needed to find the solution by Cholesky's method can be further reduced [10].

EXAMPLE 3-3 Suppose for purposes of comparison it is desired to solve the same linear equations as used in Example 3-1:

$$\begin{aligned}
-2.0x_1 + 1.1x_2 - 2.0x_3 - 1.8x_4 &= 1.0 \\
3.2x_1 + 2.1x_2 + 3.2x_3 + 2.2x_4 &= 1.0 \\
3.4x_1 + 2.3x_2 + 4.1x_3 + 3.2x_4 &= 6.0 \\
2.6x_1 + 1.1x_2 - 3.2x_3 + 2.4x_4 &= -7.0
\end{aligned}$$

The method to be used will be that of Cholesky with partial pivoting. A BASIC program that implements this problem is shown next. As before, the array A consisting of 4 rows and 5 columns is loaded with the coefficient matrix and the right sides of the preceding equation system. Notice that the working array B of size equivalent to A is not required for this method. The subroutine starting at line 3000 implements Cholesky's method with partial pivoting. The end of this subroutine is devoted to solving for the x_i values using back substitution.

```
1000  REM **********************
1010  REM *THIS PROGRAM FINDS   *
1020  REM *THE SOLUTION TO A    *
1030  REM *SET OF SIMULTANEOUS *
1040  REM *LINEAR ALGEBRAIC     *
1050  REM *EQUATIONS BY THE     *
1060  REM *CHOLESKY METHOD      *
1070  REM *USING PARTIAL        *
1080  REM *PIVOTING.            *
1090  REM **********************
1100 :
1110 :
1120  REM **SET UP THE ARRAY**
1130 :
1140  DIM A(5,6),X(6)
1150 NROW = 4:NCOL = 5
1160  FOR J = 1 TO NROW
1170  FOR I = 1 TO NCOL
1180  READ A(J,I)
1190  NEXT I
1200  NEXT J
1210  DATA  -2.0,1.1,-2.0,-1.8,1
1220  DATA  3.2,2.1,3.2,2.2,1
1230  DATA  3.4,2.3,4.1,3.2,6
1240  DATA  2.6,1.1,-3.2,2.4,-7
1250 :
1260 :
1270  REM ** APPLY CHOLESKY **
1280  REM ** METHOD         **
1290 :
1300  GOSUB 3000
1310 :
1320 :
1330  REM **WRITE THE ANSWERS**
1340 :
1350  PRINT "-----------------"
1360  PRINT "ANSWERS ARE"
1370  PRINT "-----------------"
1380  FOR J = 1 TO NROW
1390  PRINT "X(";J")=";X(J)
1400  NEXT J
1410  PRINT "-----------------"
1420  PRINT
1430 :
1440  END
1450 :
1460 :
1470 :
3000  REM **********************
3010  REM * THIS SUBROUTINE     *
3020  REM * APPLIES CHOLESKY'S *
3030  REM * METHOD TO FIND THE *
3040  REM * SOLUTION TO A SET   *
3050  REM * OF SIMULTANEOUS     *
3060  REM * LINEAR ALGEBRAIC    *
3070  REM * EQUATIONS USING     *
3080  REM * PARTIAL PIVOTING.   *
3090  REM *                     *
3100  REM *   PARAMETERS:       *
3110  REM *                     *
3120  REM *   A - THE EQUATION *
3130  REM *       MATRIX OF     *
3140  REM *       DIMENSION     *
3150  REM *       (NROW+1,      *
3160  REM *         NCOL+1).    *
3170  REM *                     *
3180  REM *   X - A VECTOR CON-*
3190  REM *       TAINING THE   *
3200  REM *       ANSWERS ON    *
3210  REM *       COMPLETION.   *
3220  REM *       IT IS DIMEN- *
3230  REM *       SIONED TO     *
3240  REM *       (NCOL+1).     *
3250  REM *                     *
3260  REM * NROW- THE NUMBER OF*
3270  REM *       ROWS IN THE   *
3280  REM *       EQUATION      *
3290  REM *       SYSTEM.       *
3300  REM *                     *
3310  REM * NCOL- THE NUMBER OF*
3320  REM *       COLUMNS IN    *
3330  REM *       THE EQUATION *
3340  REM *       SYSTEM.       *
3350  REM *                     *
3360  REM **********************
3370 :
3380 :
3390  REM **USE LARGEST    **
3400  REM **PIVOT ELEMENT **
3410 :
```

```
3420  FOR K = 1 TO NROW
3430 :
3440 :
3450  REM **FIND LARGEST PIVOT**
3460 :
3470 PIVOT = A(K,K):IL = K
3480  FOR L = K + 1 TO NROW
3490  IF  ABS (A(L,K)) <  ABS (PI
     VOT) THEN  GOTO 3520
3500 PIVOT = A(L,K)
3510 IL = L
3520  NEXT L
3530  IF IL = K THEN  GOTO 3640
3540 :
3550 :
3560  REM **TRADE ROWS TO GET**
3570  REM **LARGEST PIVOT    **
3580 :
3590  FOR LL = 1 TO NCOL
3600 TEMP = A(K,LL)
3610 A(K,LL) = A(IL,LL)
3620 A(IL,LL) = TEMP
3630  NEXT LL
3640  NEXT K
3650 :
3660 :
3670  REM **CALCULATE FIRST ROW
3680 :
3690  FOR J = 2 TO NCOL
3700 A(1,J) = A(1,J) / A(1,1)
3710  NEXT J
3720 :
3730 :
3740  REM **DO THE ROWS & COLS
3750 :
3760  FOR L = 2 TO NROW
3770 :
3780 :
3790  REM **DO THE LTH COLUMN
3800 :
3810  FOR I = L TO NROW
3820 SUM = 0.
3830  FOR K = 1 TO L - 1
3840 SUM = SUM + A(I,K) * A(K,L)
3850  NEXT K
3860 A(I,L) = A(I,L) - SUM
3870  NEXT I
3880 :
3890 :
3900  REM **DO THE LTH ROW**
3910 :
3920  FOR J = L + 1 TO NCOL
3930 SUM = 0.
3940  FOR K = 1 TO L - 1
3950 SUM = SUM + A(L,K) * A(K,J)
3960  NEXT K
3970 A(L,J) = (A(L,J) - SUM) / A(
     L,L)
3980  NEXT J
3990  NEXT L
4000 :
4010 :
4020  REM **GET X(I) VALUES BY**
4030  REM **BACK SUBSTITUTION **
4040 :
4050 X(NROW) = A(NROW,NCOL)
4060  FOR M = 1 TO NROW - 1
4070 I = NROW - M
4080 SUM = 0.
4090  FOR J = I + 1 TO NROW
4100 SUM = SUM + A(I,J) * X(J)
4110  NEXT J
4120 X(I) = A(I,NCOL) - SUM
4130  NEXT M
4140  RETURN
```

The output from this program is as follows:

```
------------------
ANSWERS ARE
------------------
X(1)=-5.20995659
X(2)=1.4262529
X(3)=1.64909867
X(4)=4.27255194
------------------
```

This program required less than 2 seconds to run on an Apple II computer and is thus about twice as fast as the Gauss–Jordan method used in Example 3-1. An Apple II computer with 48K of memory could handle up to 78 simultaneous equations with this program.

3.5 ITERATIVE METHODS FOR SIMULTANEOUS LINEAR EQUATIONS

Although direct methods are normally quite efficient at providing a solution, they tend to become less efficient than indirect methods when they are applied to sparse matrices. Sparse matrices arise in simultaneous systems where most of the equations involve only a few of the unknowns. For such systems, the amount of storage space required for iterative solution on a computer will be far less than would be required for a direct method. Thus, for reasons of both computational efficiency and storage requirements, iterative solution methods are attractive for solution by the small computer. Several iterative methods for simultaneous linear equations will now be discussed.

Iterative schemes for simultaneous linear equations are based on a formulation of the equations in which each of the n variables stands alone on the left side of one of the n equations. This form will be

$$\begin{aligned}
x_1 &= b_{1,n}x_n + b_{1,n-1}x_{n-1} + \cdots + b_{1,1}x_1 + b_{1,0} \\
x_2 &= b_{2,n}x_n + b_{2,n-1}x_{n-1} + \cdots + b_{2,1}x_1 + b_{2,0} \\
&\vdots \\
x_n &= b_{n,n}x_n + b_{n,n-1}x_{n-1} + \cdots + b_{n,2}x_2 + b_{n,1}x_1 + b_{n,0}
\end{aligned}$$

The iterative techniques utilizing this formulation are the Jacobi method, the Gauss–Seidel method, and the method of successive over-relaxation. These methods are based on the successive improvement of initial guesses for the solution. The closer the initial approximations are to the actual solution, the fewer will be the number of iterations required to achieve a solution by any of these methods.

3.6 JACOBI METHOD

In the Jacobi method, the initial guesses are used to generate new values for x_1 through x_n using the equations shown previously. If each of these new values is sufficiently close to the initial values, the process is terminated. If not, the new values replace the previous values, and the process is repeated until convergence is obtained or until it is clear that the process will never converge. In this method the replacement of solution values for all variables is done at the same time. Thus the method is sometimes called the *method of simultaneous replacement*. A primary disadvantage of this method is that a complete duplicate array of solution values must be generated before the replacement step can take place. Not only does this requirement represent an inefficient utilization of computer space, but it also slows the convergence to a solution. For these reasons the other two iterative methods are more frequently used.

3.7 GAUSS-SEIDEL METHOD

One characteristic of the method of simultaneous displacements that makes it converge so slowly is the fact that the impact of an improvement is not exploited until every unknown value has been recalculated (i.e., when the replacement is made). The method of Gauss–Seidel, also known as the *method of successive replacement* or the Liebman procedure, makes use of an improvement as soon as it is available. Thus an improvement found for x_1 is used immediately in the calculation for x_2. The new values of x_1 and x_2 are then used to calculate x_3, and so on. Since this method uses new information as soon as it is available, it can give considerable improvement in the rate of convergence and also provides considerable saving in the required computational storage space.

3.8 SUCCESSIVE OVERRELAXATION

The basis for relaxation methods is a technique that successively reduces the residual for each unknown value. The residual is the amount by which an unknown value differs from the correct solution. The method of overrelaxation is based on a linear extrapolation using two successive replacement steps. In this sense the method of successive overrelaxation can be viewed as an extension of the Gauss–Seidel method. In the method of overrelaxation, the new values computed for each variable will be

$$x_i^{(n+1)} = x_i^{(n)} + w(\overline{x}_i^{(n+1)} - x_i^{(n)})$$

where $\overline{x}_i^{(n+1)}$ is the Gauss–Seidel improved value and w is the relaxation factor, which satisfies

$$1 \leqslant w \leqslant 2$$

If $w = 1$, this technique reduces to the Gauss–Seidel method. The selection of the value w will influence the rate of convergence. Although it is beyond the scope of this text to provide a procedure for the computation of an optimum value for w, the interested reader may consult

Ames [1] for an explanation of the procedure. A useful alternative to the computation of a value for w is to choose an arbitrary value for w on the interval from 1 to 2 and to observe how the convergence progresses.

A primary disadvantage of these three iterative techniques is that they do not always converge to a solution. It can be shown that the likelihood of convergence can be improved if the iterative equations are formed so that *dominant* terms are used to solve for individual unknowns. This means that the term used to solve for an unknown should have the largest possible coefficient relative to the other unknowns in that equation. The process of selecting the dominant term for iterative methods is in many respects similar to the process of selecting the largest pivot element for direct methods and may require some manipulation or exchange of equations. Fortunately, many engineering equation systems that are sparse exhibit dominant coefficient behavior.

EXAMPLE 3-4 Suppose that it is desired to find the solution to the equation system

$$\begin{bmatrix} 2 & 1 & 0 & 0 & 0 & 0 \\ 1 & 2 & 1 & 0 & 0 & 0 \\ 0 & 1 & 2 & 1 & 0 & 0 \\ 0 & 0 & 1 & 2 & 1 & 0 \\ 0 & 0 & 0 & 1 & 2 & 1 \\ 0 & 0 & 0 & 0 & 1 & 2 \end{bmatrix} \begin{bmatrix} x_1 \\ x_2 \\ x_3 \\ x_4 \\ x_5 \\ x_6 \end{bmatrix} = \begin{bmatrix} 3.1 \\ 5.1 \\ 5.1 \\ 5.1 \\ 5.1 \\ 2.9 \end{bmatrix}$$

Since the system is sparse, an iterative method will be used to solve the system. From this equation system it can be seen that the Gauss–Seidel iteration steps will be

$$x_1 = \frac{3.1 - x_2}{2}$$

$$x_i = \frac{5.1 - x_{i-1} - x_{i+1}}{2}, \qquad \text{for} \quad i = 2, 3, \ldots, 5$$

$$x_6 = \frac{2.9 - x_5}{2}$$

These relationships are included in the following BASIC program. This program uses an initial guess of $x_i = 1$ for

$i = 1, 2, \ldots, 6$ and is written to accommodate different choices for w in the overrelaxation relationship. From this program it is obvious that the amount of working storage required to accommodate the iteration process is roughly two times the dimensional size of the coefficient matrix. To do this same problem with the Gauss–Jordan elimination method would require working storage on the order of twice the square of the dimensional size of the coefficient matrix. Thus the space saving becomes obvious for large systems that are sparse.

```
1000  REM **********************
1010  REM *THIS PROGRAM FINDS   *
1020  REM *THE SOLUTION TO A    *
1030  REM *SET OF SIMULTANEOUS *
1040  REM *LINEAR ALGEBRAIC     *
1050  REM *EQUATIONS BY         *
1060  REM *ITERATION USING THE *
1070  REM *METHOD OF SUCCESSIVE*
1080  REM *OVER RELAXATION.     *
1090  REM **********************
1100 :
1110 :
1120  REM **SET INITIAL VALUES**
1130 :
1140  DIM X(7),Y(7)
1150  FOR I = 1 TO 6
1160 X(I) = 1
1170  NEXT I
1180 :
1190 :
1200  REM *BEGIN ITERATIVE**
1210  REM *PROCESS          **
1220 :
1230 NCT = 0
1240 W = 1.
1250  PRINT "STARTING VALUES"
1260  GOSUB 1690
1270  FOR I = 1 TO 6
1280 Y(I) = X(I)
1290  NEXT I
1300 NCT = NCT + 1
1310  IF NCT > 50 THEN  GOTO 1600

1320 XBAR = (3.1 - X(2)) / 2.
1330 X(1) = X(1) + W * (XBAR - X(
     1))
1340  FOR I = 2 TO 5
1350 XBAR = (5.1 - X(I - 1) - X(I
     + 1)) / 2.
1360 X(I) = X(I) + W * (XBAR - X(
     I))
1370  NEXT I
1380 XBAR = (2.9 - X(5)) / 2.
1390 X(6) = X(6) + W * (XBAR - X(
     6))
1400  FOR I = 1 TO 6
1410  IF  ABS (Y(I) - X(I)) > 0.0
     001 THEN  GOTO 1270
1420  NEXT I
1430 :
1440 :
1450  REM **CONVERGENCE   **
1460  REM **PRINT SUMMARY**
1470 :
1480  PRINT "W=";W
1490  PRINT "CONVERGENCE REACHED"

1500  PRINT NCT;" ITERATIONS REQU
     IRED"
1510  PRINT "FINAL VALUES ARE"
1520  GOSUB 1690
1530  END
1540 :
1550 :
1560  REM **NONCONVERGENCE **
1570  REM **PRINT SUMMARY   **
1580 :
1590  PRINT "W=";W
1600  PRINT "NO CONVERGENCE "
1610  PRINT "AFTER 50 ITERATIONS"
1620  PRINT "LAST VALUES USED "
1630  GOSUB 1690
1640  END
1650 :
1660 :
1670  REM **PRINT SUBROUTINE**
1680 :
1690  PRINT "-------------------
     "
1700  FOR I = 1 TO 6
1710  PRINT "X(";I;")=";X(I)
1720  NEXT I
1730  PRINT "-------------------
     "
1740  PRINT : PRINT : PRINT
1750  RETURN
```

The output of this program is

```
STARTING VALUES
--------------------
X(1)=1
X(2)=1
X(3)=1
X(4)=1
X(5)=1
X(6)=1
--------------------


W=1
CONVERGENCE REACHED
33 ITERATIONS REQUIRED
FINAL VALUES ARE
--------------------
X(1)=.785496963
X(2)=1.52892425
X(3)=1.25674647
X(4)=1.05749999
X(5)=1.72831339
X(6)=.585843305
--------------------
```

This program required less than 10 seconds to complete the 33 iterations on an Apple II computer. If the relaxation factor w is changed to 1.4, the program uses only 14 iterations and the run time is reduced to 5 seconds. For this case, the output is

```
STARTING VALUES
--------------------
X(1)=1
X(2)=1
X(3)=1
X(4)=1
X(5)=1
X(6)=1
--------------------


W=1.4
CONVERGENCE REACHED
14 ITERATIONS REQUIRED
FINAL VALUES ARE
--------------------
X(1)=.785686222
X(2)=1.52859928
X(3)=1.25711714
X(4)=1.05715737
X(5)=1.7285651
X(6)=.585716582
--------------------
```

3.9 SOLUTION OF NONLINEAR SIMULTANEOUS ALGEBRAIC EQUATIONS

Unlike linear systems, nonlinear systems cannot be solved by direct methods because none exist. Thus the solution of nonlinear systems is always accomplished by iterative methods.

The most general formulation for nonlinear algebraic systems can be stated in the following form. Given n functions f_i in terms of n unknown variables x_i,

$$\begin{aligned} f_1(x_1, x_2, \ldots, x_n) &= 0 \\ f_2(x_1, x_2, \cdots, x_n) &= 0 \\ &\vdots \\ f_n(x_1, x_2, \ldots, x_n) &= 0 \end{aligned}$$

find the solutions. In the following sections we will deal with the common solution methods and their limitations.

3.10 DIRECT ITERATION

The method of direct iteration for the solution of nonlinear algebraic equations is actually an extension of the method of direct iteration for single equations. It is based on the assumption that the equation system can be manipulated into the form

$$\begin{aligned} x_1^* &= g_1(x_1, x_2, x_3, \ldots, x_n) \\ x_2^* &= g_2(x_1^*, x_2, x_3, \ldots, x_n) \\ &\vdots \\ x_n^* &= g_n(x_1^*, x_2^*, x_3^*, \ldots, x_n) \end{aligned}$$

The procedure for solution extraction is illustrated in Figure 3-1 and proceeds as follows. Using initial x_i guesses and any new values x_{i+1}^*, new values $x_1^*, x_2^*, \ldots, x_n^*$ are computed from the g_i equations. The x_i^* values are compared with the previous x_i values to see if the change is sufficiently small. If the change in every variable is small enough, the process is terminated. If the change in any variable value is too large, the process is repreated using the x_i^* values as new starting values. While this solution method is straightforward, it is not without its pitfalls. For example, if the initial guesses are not sufficiently close to the true solution, the process will fail to converge. The space domain within which an initial guess will converge to a solution is called the *domain of*

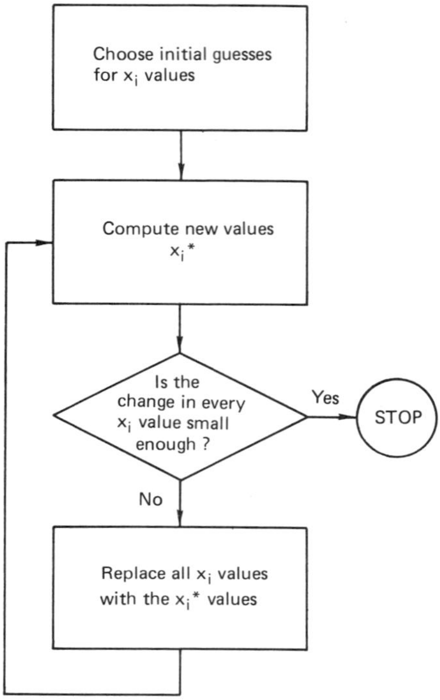

Figure 3-1 Logic flow diagram for the method of direct iteration.

convergence. Any initial guesses outside this domain will not lead to a solution. It is an unfortunate reality for nonlinear algebraic systems that, as the number of equations (and thus the number of unknowns), goes up, the size of the domain of convergence gets smaller. Indeed, for extremely large systems the initial guesses must be very close to the actual solution values in order to achieve convergence. Although there do exist some methods for improving the convergence difficulties in problems of this type, there is no substitute for good judgment in the selection process of initial values to start the iteration process.

3.11 NEWTON'S ITERATION METHOD

Newton's iteration method is, by far, the most commonly used method for the solution of systems of nonlinear algebraic equations. Its popularity is due to the fact that it has better convergence properties than does the method of direct iteration. The basis for Newton's iteration method is a Taylor expansion of each of the n equations:

$$f_1(x_1 + \Delta x_1, \ldots, x_n + \Delta x_n) = f_1(x_1, \ldots, x_n) + \Delta x_1 \frac{\partial f_1}{\partial x_1} + \cdots + \Delta x_n \frac{\partial f_1}{\partial x_n} + \text{higher-order terms}$$

$$\vdots$$

$$f_n(x_1 + \Delta x_1, \ldots, x_n + \Delta x_n) = f_n(x_1, \ldots, x_n) + \Delta x_1 \frac{\partial f_n}{\partial x_1} + \cdots + \Delta x_n \frac{\partial f_n}{\partial x_n} + \text{higher-order terms}$$

If the Δx_i changes in the variable values bring the functions f_j close to a root, it will be assumed that the left sides of these equations are zero. Thus the problem reduces to that of finding the changes Δx_i that achieve the goal. If the higher-order terms are dropped, the problem becomes one of finding the roots of the linear system:

$$\begin{bmatrix} \frac{\partial f_1}{\partial x_1} & \frac{\partial f_1}{\partial x_2} & \cdots & \frac{\partial f_1}{\partial x_n} \\ \vdots & & & \\ \frac{\partial f_n}{\partial x_n} & & \cdots & \frac{\partial f_n}{\partial x_n} \end{bmatrix} \begin{bmatrix} \Delta x_1 \\ \Delta x_2 \\ \vdots \\ \Delta x_n \end{bmatrix} = \begin{bmatrix} -f_1 \\ -f_2 \\ \vdots \\ -f_n \end{bmatrix}$$

In this system the partial derivative matrix and the vector on the right side can each be evaluated at any approximate set of solution values. Once the x_i values are known, they may be applied as corrections to the initial approximations:

$$x_1 = x_1 + \Delta x_1$$
$$\vdots$$
$$x_n = x_n + \Delta x_n$$

If all correction factors are sufficiently small, the process is terminated. If not, the new values are used as root approximations, and the process is repeated until a solution is found or until it is obvious that no solution can be achieved. The logic flow diagram for this procedure is shown in Figure 3-2. Care should be exercised in the test for convergence. If the individual root values x_i are of greatly different magni-

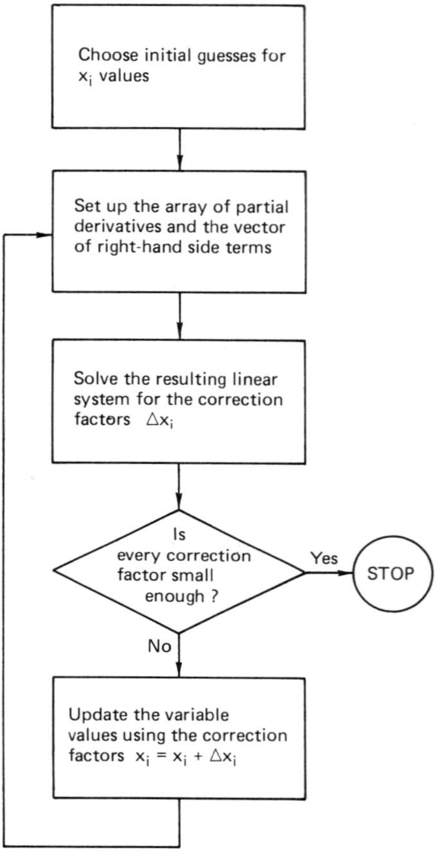

Figure 3-2 Logic flow diagram for the Newton iteration method.

tudes, a check requirement such as

$$|\Delta x_i| \leqslant 0.0001, \qquad i = 1, \ldots, n$$

may place unfair limitations on the larger x_i values. When this situation occurs, it is better to test a normalized correction factor:

$$\left|\frac{\Delta x_i}{x_i}\right| \leqslant 0.0001, \qquad i = 1, \ldots, n$$

Although the Newton iteration method is the most preferable of the iterative methods, it still can have convergence problems. The size of the domain of convergence is roughly inversely related to the degree and number of equations.

EXAMPLE 3-5 Suppose that it is desired to solve the following system of four equations in four unknowns:

$$x_1 + 2x_2 + x_3 + 4x_4 = 20.700$$
$$x_1^2 + 2x_1x_2 + x_4^3 = 15.880$$
$$x_1^3 + x_3^2 + x_4 = 21.218$$
$$3x_2 + x_3x_4 = 21.100$$

The method to be used will be that of Newton iteration. To implement this method, it is necessary to find the partial derivatives of each of the equations in the null form. This matrix of partial derivatives is then used to compute correction factors for initial guesses for the four x_i values. The calculation of correction factors requires the solution of four linear equations in four unknowns. The following computer program implements the solution of this example problem. This program utilizes the Gauss–Jordan elimination portion of the program in Example 3-1.

```
1000  REM **********************
1010  REM * THIS PROGRAM FINDS *
1020  REM * THE ROOTS TO A SET *
1030  REM * OF NONLINEAR,      *
1040  REM * SIMULTANEOUS,      *
1050  REM * ALGEBRAIC EQNS.    *
1060  REM * USING THE METHOD   *
1070  REM * OF NEWTON ITERATION*
1080  REM **********************
1090 :
1100 :
1110  DIM A(5,6),B(5,6),X(5),G$(5
     )
1120 NCOL = 5:NROW = 4
1130 :
1140 :
1150  REM **SET INITIAL GUESSES*
1160 :
1170 X(1) = 1:X(2) = 1:X(3) = 1:X
     (4) = 1
1180 :
1190 :
1200  PRINT "--------------------
     ------------------"
1210  PRINT "  X(1)"; TAB( 13);"X
     (2)"; TAB( 23);"X(3)"; TAB(
     33);"X(4)"
1220  PRINT "--------------------
     ------------------"
1230  GOSUB 3000
1240  REM **ITERATE FOR SOLUTION*

1250 :
1260  FOR IA = 1 TO 50
1270 A(1,1) = 1:A(1,2) = 2:A(1,3)
      = 1
1280 A(1,4) = 4:A(2,3) = 0:A(3,2)
      = 0
1290 A(3,4) = 1:A(4,1) = 0:A(4,2)
      = 3.
1300 A(2,1) = 2 * X(1) + 2 * X(2)

1310 A(2,2) = 2 * X(1)
1320 A(2,4) = 3 * X(4) ^ 2
1330 A(3,1) = 3 * X(1) ^ 2
1340 A(3,3) = 2 * X(3)
1350 A(4,3) = X(4):A(4,4) = X(3)
1360 A(1,5) =  - 1 * X(1) - 2 * X
     (2) - X(3) - 4 * X(4) + 20.7

1370 A(2,5) = 15.88 - X(1) ^ 2 -
     2 * X(1) * X(2) - X(4) ^ 3
1380 A(3,5) = 21.218 - X(1) ^ 3 -
     X(3) ^ 2 - X(4)
1390 A(4,5) =  - 3 * X(2) - X(3) *
     X(4) + 21.1
1400  FOR IJ = 1 TO 4
1410  IF  ABS (A(IJ,5)) > 0.00001
      THEN  GOTO 1440
1420  NEXT IJ
```

```
1430  GOTO 1560
1440  GOSUB 2000
1450 :
1460  REM **UPDATE X VALUES**
1470  FOR IJ = 1 TO 4
1480 X(IJ) = X(IJ) + A(IJ,5)
1490  NEXT IJ
1500  GOSUB 3000
1510  NEXT
1520  PRINT "NO CONVERGENCE"
1530  PRINT "AFTER 50"
1540  PRINT "ITERATIONS"
1550  END
1560  PRINT "CONVERGENCE OBTAINED
   "
1570  END
1580 :
1590 :
2000  REM *********************
2010  REM * THIS SUBROUTINE    *
2020  REM * APPLIES THE METHOD *
2030  REM * OF GAUSS-JORDAN    *
2040  REM * ELIMINATION TO A   *
2050  REM * MATRIX USING       *
2060  REM * PARTIAL PIVOTING.  *
2070  REM *                    *
2080  REM *   PARAMETERS:      *
2090  REM *                    *
2100  REM *    A - THE ORIGINAL*
2110  REM *        AUGMENTED   *
2120  REM *        MATRIX      *
2130  REM *        DIMENSIONED *
2140  REM *        (NROW+1,    *
2150  REM *           NCOL+1). *
2160  REM *                    *
2170  REM *    B - A WORKING   *
2180  REM *        MATRIX OF   *
2190  REM *        SIZE SIMILAR*
2200  REM *        TO "A."     *
2210  REM *                    *
2220  REM *  NROW- THE NUMBER  *
2230  REM *        OF ROWS IN  *
2240  REM *        THE A MATRIX*
2250  REM *                    *
2260  REM *  NCOL- THE NUMBER  *
2270  REM *        OF COLUMNS  *
2280  REM *        IN THE "A"  *
2290  REM *        MATRIX.     *
2300  REM *                    *
2310  REM *********************
2320 :
2330 :
2340  REM **DO THE PROBLEM  **
2350  REM **IN STAGES       **
2360 :
2370  FOR K = 1 TO NROW
2380 :
2390 :
2400  REM **FIND LARGEST PIVOT**
2410 :
2420 PIVOT = A(K,K):IL = K
2430  FOR L = K + 1 TO NROW
2440  IF  ABS (A(L,K)) < ABS (PI
     VOT) THEN  GOTO 2470
2450 PIVOT = A(L,K)
2460 IL = L
2470  NEXT L
2480 :
2490 :
2500  REM **TRADE ROWS TO GET**
2510  REM **LARGEST PIVOT    **
2520 :
2530  FOR LL = 1 TO NCOL
2540 TEMP = A(K,LL)
2550 A(K,LL) = A(IL,LL)
2560 A(IL,LL) = TEMP
2570  NEXT LL
2580 :
2590 :
2600  REM **NORMALIZE PIVOT ROW*
2610 :
2620  FOR J = 1 TO NCOL
2630 B(K,J) = A(K,J) / PIVOT
2640  NEXT J
2650 :
2660  REM **DO GAUSS-JORDAN**
2670  REM * ELIMINATION STEP*
2680 :
2690  FOR I = 1 TO NROW
2700  IF I = K GOTO 2740
2710  FOR J = 1 TO NCOL
2720 B(I,J) = A(I,J) - A(I,K) * B
     (K,J)
2730  NEXT J
2740  NEXT I
2750 :
2760 :
2770  REM **UPDATE 'A' MATRIX**
2780  REM **WITH 'B' MATRIX  **
2790 :
2800  FOR I = 1 TO NROW
2810  FOR J = 1 TO NCOL
2820 A(I,J) = B(I,J)
2830  NEXT J
2840  NEXT I
2850  NEXT K
2860  RETURN
2870 :
2880 :
3000  REM *********************
3010  REM * THIS SUBROUTINE    *
3020  REM * PRINTS THE RESULTS *
3030  REM *********************
3040 :
3050 :
3060  GOSUB 3120
```

```
3070  PRINT G$(1); TAB( 11);G$(2)
     ; TAB( 21);G$(3); TAB( 31);G
     $(4)
3080 :
3090  RETURN
3100 :
3110 :
3120  REM ********************
3130  REM * THIS SUBROUTINE    *
3140  REM * SORTS SCIENTIFIC   *
3150  REM * NOTATION FOR PRINT*
3160  REM * CLARITY.           *
3170  REM ********************
3180 :
3190 :
3200  FOR I = 1 TO 4
3210  IF  ABS (X(I)) > 0.01 THEN
      GOTO 3240
3220 G$(I) =  LEFT$ ( STR$ (X(I))
     ,2) +  RIGHT$ ( STR$ (X(I)),
     4)
3230  GOTO 3260
3240 G$(I) =  STR$ (X(I))
3250 G$(I) =  LEFT$ (G$(I),8)
3260  NEXT I
3270  RETURN
```

The output of this program is

```
-----------------------------------------
  X(1)        X(2)        X(3)        X(4)
-----------------------------------------
1          1          1          1
2.750367   4.676298   7.895793   .1753103
1.344849   5.297124   5.949349   .7028880
1.477496   3.843727   4.341854   1.798298
1.542659   6.243380   4.120361   .6375545
1.236366   5.727357   4.343621   .9163241
1.202389   5.598621   4.299491   1.000219
1.200000   5.600000   4.300002   .9999992
1.2        5.600000   4.3        .9999999

CONVERGENCE OBTAINED
```

This program required less than 40 seconds to complete the eight iterations on an Apple II computer.

3.12 PARAMETER PERTURBATION PROCEDURE

The parameter perturbation procedure [1] is an algorithm for enabling an iterative solution method to achieve a root to a set of simultaneous nonlinear equations. It is a procedure that operates independently of the need for a "good" initial approximation. The procedure is implemented as follows. First, given a set of equations,

$$f_j(x_i) = 0, \qquad i = 1, \ldots, n$$
$$j = 1, \ldots, n$$

Consider another set of equations,

$$g_j(x_i) = 0, \qquad i = 1, \ldots, n$$
$$j = 1, \ldots, n$$

for which a solution set is known. The $g_j = 0$ equations are "deformed" into the equations $f_j = 0$ by means of a finite number N of successive small increments in the parameters:

$$g_j^{(k)}(x_i) = g_j^{(k-1)}(x_i) + (f_j(x_i) - g_j^{(k-1)}(x_i))\,\frac{k}{N}, \qquad k = 1, \ldots, N$$

The initial solution set $x_i^{(0)}$ to the initial set of equations $g_j^{(0)}(x_i)$ can be used as an initial guess for the iterative solution of the $g_j^{(1)}(x_i)$ set. Since this set is only slightly different from the $g_j^{(0)}(x_i)$ set, convergence is likely. As the procedure progresses, the $x_i^{(k-1)}$ solution becomes the initial guess to obtain the $x_i^{(k)}$ solution. In the end, when $k = N$, the system is equivalent to the original set of equations. Since it may require $N = 10$ or $N = 100$ steps to deform the known problem into the desired problem, the procedure can require considerable computer time to implement. Fortunately, if the step sizes are small, convergence at any single step can often be accomplished with only a few iterations.

The parameter perturbation procedure has been shown to be especially useful for the solution of equation systems encountered in the synthesis of mechanical linkages. In such cases any arbitrary mechanism of the type required can serve to generate the starting system $g_j^{(0)}(x_i) = 0$.

3.13 CONSIDERATIONS IN THE SELECTION OF AN ALGORITHM FOR THE SMALL COMPUTER

Although it is impossible to state universal rules that will guide the user in selecting the best method for finding the solutions for a particular equation system, a few basic guidelines do exist. These are as follows:

1. **Consider the nature of the problem and its solutions.** If the equation system to be solved is linear, the user can be assured that there will be one and only one solution. For most reasonably sized systems, a direct method such as Cholesky's method is best to use. This method guarantees a solution as long as the equations in the system are linearly independent. If the linear equation system is sparse, the solution may be found using an iterative approach such as the method of successive overrelaxation. This method does require that the user know some information about the solution, since the ability of the system to converge rapidly to a solution will depend strongly on the nearness of the initial guesses to the actual solution. If the user does not have any information to help with the

initial selection of iteration values, it is probably best to choose a direct method.

If the equation system to be solved is nonlinear, there may be more than one solution. For most nonlinear equation systems, the best technique to use is the Newton iteration method. The only disadvantage to this method is that it requires the user to find the partial derivatives of the equations with respect to the variables. Only if it is not possible to find these derivatives should another iterative method be used. For the Newton iteration method, the ability to converge to a solution will depend on how near the initial guesses are made to the actual solution. The domain within which convergence will occur is known as the domain of convergence. The size of the domain of the convergence decreases as the number of equations increases. The method of parameter perturbation will often help to achieve a solution when the domain of convergence becomes quite small; however, even this method will sometimes not guarantee convergence. For such situations, the user can only try different starting values in the hope that one choice will lead to a converging solution.

2. **Consider the computer space and run time required.** There is frequently a tradeoff between computer run time and computer storage space that will determine which method must be used for equation systems on a small computer. For example, the iterative methods for linear systems usually take far less storage space than do the direct methods. On the other hand, if little is known about the solution, an iterative method may take a very long time to converge or may even fail to converge altogether.
3. **Consider the intermediate outputs of the computer.** Even though the process of outputting information to a CRT screen or to a printer may slow down an iterative solution somewhat, this intermediate information concerning the progress of the technique toward achieving a solution can often give the user considerable information as to whether this solution will ultimately converge or is failing to converge. For some types of iterative problems, the solution may appear to be oscillating rather than converging. This situation can be identified quickly by a visual inspection of intermediate iteration values. When this situation occurs, the user may need to start with a different set of initial values or may even need to alter the solution technique.
4. **Consider the accuracy required.** Care should be exercised in establishing the criteria for measuring convergence of any iterative process. For example, in the Newton iteration method it is best to measure the degree of convergence by looking at the relative size of

the corrections made to the iterated values. For example, if the individual root values are of greatly different magnitudes, a check should be made on a normalized correction factor rather than on the actual correction factor itself.

REFERENCES

1. AMES, W.F., *Numerical Methods for Partial Differential Equations*, Barnes & Noble Books, New York, 1969.
2. FREUDENSTEIN, F., and B. ROTH, "Numerical Solution of Systems of Nonlinear Equations." *Journal of the Association for Computing Machines*, 10, 4 (1963), 550-556.
3. GROVE, WENDELL E., *Brief Numerical Methods*, Prentice-Hall, Inc., Englewood Cliffs, N.J., 1966.
4. HORNBECK, R.W., *Numerical Methods*, Quantum Publishers, Inc., New York, 1975.
5. LA FARA, ROBERT L., *Computer Methods for Science and Engineering*, Hayden Book Co., Inc., Rochelle Park, N.J., 1973.
6. McCALLA, THOMAS R., *Introduction to Numerical Methods and FORTRAN Programming*, John Wiley & Sons, Inc., New York, 1967.
7. PALL, GABRIEL A., *Introduction to Scientific Computing*, Appleton-Century-Crofts, Educational Division, Meredith Corp., New York, 1971.
8. RALSTON, ANTHONY, *A First Course in Numerical Analysis*, McGraw-Hill Book Co., New York, 1965.
9. RALSTON, ANTHONY, and H.S. WILF, *Mathematical Methods for Digital Computers*, John Wiley & Sons, Inc., New York, 1967.
10. SALVADORI, MARIO G., and MELVIN L. BARON, *Numerical Methods in Engineering*, Prentice-Hall, Inc., Englewood Cliffs, N.J., 1961.
11. TURNER, L.R., "Solution on Nonlinear Systems," *Annals of the New York Academy of Sciences*, Vol. 86, 1960, pp. 817-827.
12. WILLIAMS, P.W., *Numerical Computation*, Harper & Row, Publishers, Inc., New York, 1972.

Because of its portability, the microcomputer can be used in the laboratory environment to solve a variety of problems on-site. (*Photo courtesy of Radio Shack, a division of Tandy Corporation.*)

Eigenvalue problems

4

The analysis of some types of problems in science and engineering often leads to sets of algebraic equations that can have a unique solution only when the value of a parameter within the equations is known. This special parameter is called a *characteristic value* or *eigenvalue*. The solution associated with each eigenvalue is called its *eigenvector*. Eigenvalue problems occur in a variety of situations. In the manipulation of stress tensors, the eigenvalues identify the principal normal stresses, and the eigenvectors identify the orientations associated with these values. In the dynamic analysis of systems, the eigenvalues identify the natural frequencies of vibration, and the eigenvectors characterize the shapes of these vibration modes. In structural analysis, eigenvalues can be used to determine the critical loads for buckling or other modes of instability.

Selection of the best numerical technique for finding the eigenvalues and eigenvectors for a given problem will depend on several factors, such as the nature of the equations, the number of eigenvalues desired, and the nature of these eigenvalues. When using the small computer, the choice of solution technique will also be influenced by the speed, accuracy, and storage capacity of the device being used. In general, there are two categories of solution algorithms for eigenvalue problems. The iterative methods are quite easy to use and are well suited for finding the smallest and largest eigenvalues. The transformation methods are a bit more complex to apply but have the advantage that they identify all eigenvalues and eigenvectors.

It is the purpose of this chapter to discuss the common techniques available for solution of eigenvalue problems. Before presenting these techniques, it will be useful to review some fundamentals of matrix and vector theory upon which the eigenvalue methods are based.

4.1 FUNDAMENTALS OF EIGENVALUE PROBLEMS

The general formulation of an eigenvalue problem is

$$AX = \lambda X$$

where A is an $n \times n$ matrix. In this expression it is desired to find the n scalar values of λ and the eigenvectors X associated with each of these eigenvalues.

Although it is assumed that the reader has a basic knowledge of matrix calculus and of eigenvalue theory, a few fundamental principles are worthy of reemphasis before considering solution methods.

Fundamentals of Matrices

1. A matrix A is symmetric if $a_{ij} = a_{ji}$ $(i, j = 1, 2, \ldots, n)$. This gives rise to symmetry about the diagonal a_{kk} $(k = 1, 2, \ldots, n)$. An example of a symmetric matrix would be

$$\begin{bmatrix} 1 & 4 & 5 \\ 4 & 3 & 7 \\ 5 & 7 & 2 \end{bmatrix}$$

2. A matrix A is said to be tridiagonal if all elements are zero except the main diagonal elements and the diagonal elements immediately above and below the main diagonal. The general form of a tridiagonal matrix is

$$\begin{bmatrix} * & * & & & & & & & & \\ * & * & * & & & & 0 & & & \\ & * & * & * & & & & & & \\ & \cdot & \cdot & \cdot & \cdot & \cdot & \cdot & \cdot & \cdot & \\ & & & & & & * & * & * & \\ & & & 0 & & & & * & * & * \\ & & & & & & & & * & * \end{bmatrix}$$

The tridiagonal form is important because some transformation methods reduce a general matrix to this special form.

3. A matrix is orthogonal if

$$A^T A = I$$

where A^T represents the transpose of the A matrix and I represents the identity matrix. Clearly, the inverse of an orthogonal matrix is equivalent to its transpose.

4. Two matrices A and B are said to be similar if a nonsingular matrix P can be found such that

$$B = P^{-1} A P$$

Fundamentals of Eigenvalues

1. The n eigenvalues of any real symmetric $n \times n$ matrix will all be real. This fact is particularly useful since many engineering matrices are symmetrical.
2. If the eigenvalues of a matrix are distinct, then the eigenvectors will be orthogonal. A set of n linearly independent eigenvectors will form a base for the space being considered. This means that for the linearly independent set of eigenvectors

$$X^i, \qquad i = 1, \ldots, n$$

any arbitrary vector Y in the space can be expressed in terms of the eigenvectors. Thus

$$Y = \sum_{i=1}^{n} a_i X^i$$

3. If two matrices are similar, they have the same eigenvalues. Thus the similarity of A and B means

$$B = P^{-1}AP$$

Since

$$AX = \lambda X$$

then

$$P^{-1}AX = \lambda P^{-1}X$$

If one lets $X = PY$, then

$$P^{-1}APY = \lambda Y$$

and

$$BY = \lambda Y$$

Thus, not only do the two matrices have the same eigenvalues, but their eigenvectors are related by $X = PY$.

4. Any scalar multiple of an eigenvector of a matrix is also an eigenvector of that matrix. It is customary to normalize all eigenvectors either by dividing each element of the eigenvector by the largest element or by dividing each element by the sum of the squares of the other elements.

4.2 ITERATIVE METHODS OF SOLUTION

Perhaps the most obvious technique for the solution of eigenvalues can be found by expressing the eigenvalue problem as

$$(A - \lambda I)X = 0$$

This system will have a nonzero solution only if the determinant of $(A - \lambda I)$ is zero. This determinant will give a polynomial in λ of degree n, and the roots of this polynomial will be the eigenvalues. Thus, any of the methods presented in Chapter 2 may be applied to find these roots. Unfortunately, eigenvalue problems often have multiple roots. Since the iterative methods of Chapter 2 do not work well for this situation, it is best to utilize other iterative procedures to extract the eigenvalues.

Finding the Largest Eigenvalue by Iteration

The basic iterative method for finding the largest eigenvalue of the system

$$AX = \lambda X$$

is illustrated in Figure 4-1. The procedure begins with a trial normalized vector $X^{(0)}$. This vector is multiplied by the A matrix on the left side to obtain the product X. Next the product X is reduced to a constant (the eigenvalue) and a normalized vector $X^{(1)}$. If the vector $X^{(1)}$ reproduces the vector $X^{(0)}$, the process may be terminated. If not, the new normalized vector becomes the starting vector and the process is repeated. When the process has converged, the constant multiplier will be the correct value for the largest eigenvalue, and the normalized vector will be the corresponding eigenvector. The rate of convergence for this iterative process depends on the choice of the initial vector. A choice close to the actual eigenvector gives rapid convergence. The rate of convergence is also influenced by the ratio of the sizes of the two largest eigenvalues. When this ratio is nearly unity, the convergence rate is poor.

Finding the Smallest Eigenvalue by Iteration

In some engineering problems, it is more convenient to be able to find the smallest eigenvalue rather than the largest. This can be accomplished by premultiplying the original system by the inverse of the A

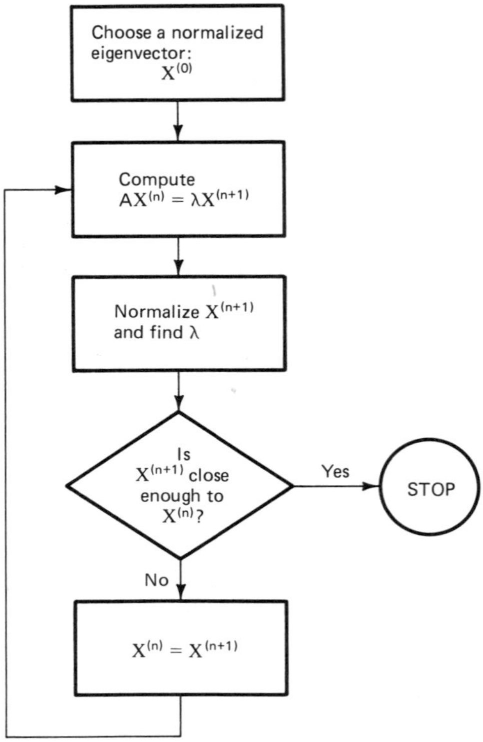

Figure 4-1 Logic flow diagram for the iterative eigenvalue method.

matrix. Thus

$$A^{-1}AX = \lambda A^{-1}X$$

If both sides are then multiplied by $1/\lambda$, the result will be

$$\frac{1}{\lambda}X = A^{-1}X$$

Clearly, this is a different eigenvalue problem, where $1/\lambda$ is the eigenvalue and A^{-1} is the matrix of interest. The maximum value of $1/\lambda$ will occur when λ is smallest. Thus the previous iterative scheme can be applied to this new system to isolate the smallest eigenvalue.

Finding Intermediate Eigenvalues by Iteration

After one has found the largest eigenvalue, it is possible to find the next largest eigenvalue by replacing the original matrix with one that possesses only the remaining eigenvalues. This process of removing the largest known eigenvalue is known as *deflation*. For an original symmetrical matrix A with known largest eigenvalue λ_1 and eigenvector X_1, the principle of orthogonality of the eigenvectors can be invoked. This means

$$X_i^T X_j = 0, \qquad \text{for } i \neq j$$
$$= 1, \qquad \text{for } i = j$$

If a new matrix A^* is formed by

$$A^* = A - \lambda_1 X_1 X_1^T$$

the eigenvalues and eigenvectors of this new matrix will be found from

$$A^* X_i = \lambda_i X_i$$

From the previous expression for A^*, it is clear that

$$A^* X_i = AX_i - \lambda_1 X_1 X_1^T X_i$$

In this expression, if $i = 1$, the orthogonality relationship requires that the right side become

$$AX_1 - \lambda_1 X_1$$

but this must be zero since it is the definition of the eigenvalues of A. Thus the λ_1 eigenvalue of the A^* matrix is zero, and all other eigenvalues of A^* are the same as for A. Thus the eigenvalues of A^* are 0, $\lambda_2, \lambda_3, \ldots, \lambda_n$ having corresponding eigenvectors $X_1, X_2, \ldots, X_n$. The larger eigenvalue λ_1 has thus been removed, and the traditional iteration method can be applied to A^* to find the next largest eigenvalue λ_2. Once λ_2 and X_2 are known, the process may be repeated using a new matrix A^{**} found from A^*, λ_2, and X_2. Although this process may appear to have considerable promise, it does have fundamental drawbacks. As each new step is performed, any errors in the eigenvectors will be passed on to the next eigenvector to cause increasing inaccuracy. For this reason the method is of questionable value for use in finding more than three eigenvalues removed from the highest or lowest. When more than this number of eigenvalues is desired, it is better to utilize similarity transformation methods. Before considering the similarity transformation methods an example of the use of the iterative method will be presented.

EXAMPLE
4-1

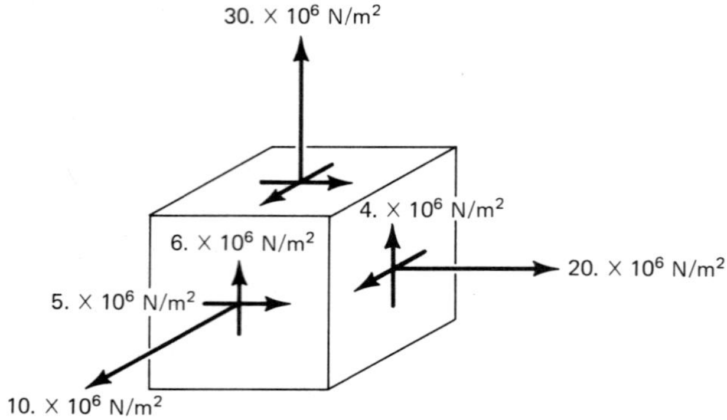

It is desired to investigate the triaxial state of stress indicated in the figure. For this elemental block, the stress matrix will be

$$\begin{bmatrix} 10. & 5. & 6. \\ 5. & 20. & 4. \\ 6. & 4. & 30. \end{bmatrix} \times 10^6 \text{ N/m}^2$$

If the maximum stress theory of failure is to be used for this investigation, it will be necessary to know the largest principal stress value. This stress corresponds to the largest eigenvalue of the stress matrix. Thus an iterative method will be used. The following computer program, which implements the iterative procedure, iterates until the change in the eigenvalue is smaller than 0.01 percent. The program utilizes subroutines to implement the matrix multiplication and the normalization of eigenvectors.

```
1000  REM **********************
1010  REM *THIS PROGRAM FINDS  *
1020  REM *THE LARGEST  EIGEN- *
1030  REM *VALUE OF A REAL     *
1040  REM *SYMMETRIC MATRIX    *
1050  REM *BY ITERATION        *
1060  REM **********************
1070 :
1080 :
1090  DIM A(4,4),X(4),R(4):N = 3
1100 :
1110  REM ***LOAD THE MATRIX
1120 A(1,1) = 10E06
1130 A(1,2) = 5E06
1140 A(1,3) = 6E06
1150 A(2,1) = A(1,2)
1160 A(2,2) = 20E06
1170 A(2,3) = 4E06
1180 A(3,1) = A(1,3)
1190 A(3,2) = A(2,3)
```

```
1200 A(3,3) = 30E06
1210 :
1220  REM ***CHOOSE INITIAL
1230  REM ***NORMALIZED EIGEN-
1240  REM ***VECTOR.
1250 X(1) = 1:X(2) = 0:X(3) = 0
1260 :
1270  GOSUB 3000
1280 :
1290  REM ***PRINT THE RESULTS
1300 :
1310  PRINT "THE EIGENVALUE IS ";
     XL;" (N/M^2)"
1320  PRINT
1330  PRINT "THE FINAL EIGENVECTO
     R IS"
1340  FOR I = 1 TO 3
1350  PRINT "X(";I;")=";X(I)
1360  NEXT I
1370  PRINT
1380  PRINT "THE NUMBER OF ITERAT
     IONS"
1390  PRINT "REQUIRED IS ";K
1400  END
1410 :
1420 :
3000  REM *********************
3010  REM * THIS SUBROUTINE    *
3020  REM * FINDS THE LARGEST  *
3030  REM * EIGENVALUE  OF  A  *
3040  REM * REAL, SYMMETRIC    *
3050  REM * MATRIX.            *
3060  REM *                    *
3070  REM * THE CORRESPONDING  *
3080  REM * EIGENVECTOR IS ALSO*
3090  REM * FOUND.             *
3100  REM *                    *
3110  REM * THE METHOD USED IS *
3120  REM * ITERATION.         *
3130  REM *                    *
3140  REM * THE PROCESS STOPS  *
3150  REM * WHEN THE CHANGE IN *
3160  REM * EIGENVALUE IS LESS *
3170  REM * THAN 0.01 PERCENT  *
3180  REM * OR 50 ITERATIONS   *
3190  REM * HAVE BEEN TRIED.   *
3200  REM *                    *
3210  REM *  PARAMETERS:       *
3220  REM *                    *
3230  REM *   N     - NUMBER OF*
3240  REM *           ROWS AND *
3250  REM *           COLUMNS  *
3260  REM *           IN MATRIX*
3270  REM *                    *
3280  REM *   A     - ORIGINAL *
3290  REM *           MATRIX   *
3300  REM *           DIMENSION*
3310  REM *           (N+1,N+1)*
3320  REM *                    *
3330  REM *                    *
3340  REM *   R     - WORKING  *
3350  REM *           ARRAY OF *
3360  REM *           SAME SIZE*
3370  REM *           AS "A."  *
3380  REM *                    *
3390  REM *   XL    - EIGEN-   *
3400  REM *           VALUE    *
3410  REM *           ANSWER.  *
3420  REM *                    *
3430  REM *   X     - VECTOR OF*
3440  REM *           EIGEN-   *
3450  REM *           VECTOR   *
3460  REM *           ANSWERS. *
3470  REM *                    *
3480  REM *   K     - NUMBER OF*
3490  REM *           ITERA-   *
3500  REM *           TIONS TO *
3510  REM *           REACH A  *
3520  REM *           SOLUTION.*
3530  REM *                    *
3540  REM *********************
3550 :
3560 XOLD = 0
3570 :
3580  FOR K = 1 TO 50
3590 :
3600  REM ***MULTIPLY A MATRIX
3610  REM ***BY X VECTOR
3620  FOR J = 1 TO N
3630 SUM = 0.
3640  FOR I = 1 TO N
3650 SUM = SUM + A(I,J) * X(I)
3660  NEXT I
3670 R(J) = SUM
3680  NEXT J
3690 :
3700  FOR JJ = 1 TO N
3710 X(JJ) = R(JJ)
3720  NEXT JJ
3730 :
3740  REM ***NORMALIZE THE
3750  REM ***EIGENVECTOR
3760  REM ***ELEMENTS WITH
3770  REM ***RESPECT TO
3780  REM ***LARGEST ELEMENT.
3790 :
3800  REM *FIND LARGEST ELEMENT
3810 XBIG = X(1)
3820  FOR I = 1 TO N
3830  IF X(I) < XBIG THEN  GOTO 3
     850
3840 XBIG = X(I)
3850  NEXT I
3860 :
3870  REM *NORMALIZE THE VECTOR
3880  FOR I = 1 TO N
```

```
3890 X(I) = X(I) / XBIG
3900  NEXT I
3910 :
3920 XL = XBIG
3930  IF  ABS (XOLD - XL) /  ABS
       (XL) <  = 0.0001 THEN  GOTO
       3960
3940 XOLD = XL
3950  NEXT K
3960  RETURN
```

The output from this program is as follows.

```
THE EIGENVALUE IS 33712189.3 (N/M^2)

THE FINAL EIGENVECTOR IS
X(1)=.340905732
X(2)=.416363575
X(3)=1

THE NUMBER OF ITERATIONS
REQUIRED IS 14
```

This program required less than 6 seconds to run on an Apple II computer. An Apple II computer with 48K of memory could handle this problem with a matrix having more than 80 rows and columns. It should be noted that larger matrices could be handled if the user could store the A matrix in a more streamlined form. This is possible since a general symmetrical matrix has only $n(n - 1)/2$ unique terms. The price paid for this space saving would be that the program would be a bit more complex and might take longer to run.

4.3 TRANSFORM METHODS OF EIGENVALUE CALCULATION

The objective in utilizing a similarity transformation method of a matrix is to obtain a new matrix having the same eigenvalues but a simpler form. The best possible simplification would be to reduce the matrix to a pure diagonal form since the eigenvalues would then be available by inspection of the diagonal elements. Unfortunately, most transformation methods do not achieve this special form. In general, we are satisfied if the matrix is reduced to a simplified form such as the tridiagonal form.

Jacobi's Method

Jacobi's method is designed to produce a diagonal form by eliminating each off-diagonal element in a systematic fashion. Unfortunately, the process requires an infinite number of steps to achieve a perfect diagonal form since the introduction of a new zero term at one point in a matrix will often introduce a nonzero element into a previous zero position. In practice, the Jacobi method can be viewed as an iterative procedure that eventually approaches the diagonal form close enough to be terminated. For a real symmetric matrix A, the computation is performed using orthogonal matrices that are real plane rotations. The computational scheme is illustrated in Figure 4-2 and proceeds as follows. A new matrix A_1 is formed from the original A matrix by means of the transformation $A_1 = P_1 A P_1^T$. The orthogonal matrix P_1 is chosen so that it introduces a zero off-diagonal element in A_1. Next a new matrix A_2 is formed from A_1 using a second transformation matrix P_2 selected so that a different off-diagonal element of A_2 will be zero. The process is continued in such a way that, at each step, the off-diagonal element of maximum magnitude is reduced to zero. The process continues until the off-diagonal element of largest magnitude is small enough to be considered zero relative to the size of the diagonal elements. The transformation matrix to accomplish this process at each stage is constructed as follows. If the a_{kl} element of A_{m-1} has maxi-

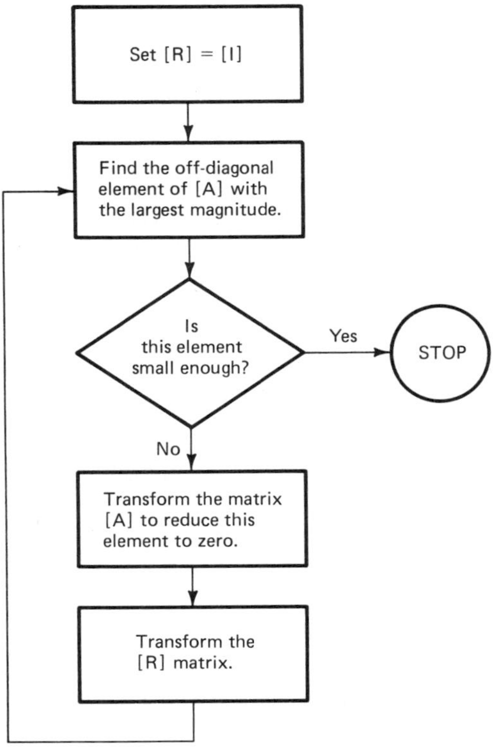

Figure 4-2 Logic flow diagram for the Jacobi eigenvalue method.

mum magnitude, then P_m corresponds to

$$
\begin{aligned}
P_{kk} &= P_{ll} = \cos\theta \\
P_{kl} &= -P_{lk} = \sin\theta \\
P_{ii} &= 1, \qquad i \neq k, l; \; P_{ij} = 0 \text{ otherwise}
\end{aligned}
$$

The A_m matrix will differ from the A_{m-1} matrix only in rows and columns k and l. To ensure that the $a_{kl}^{(m)}$ element be zero, the value of θ is selected so that

$$
\tan 2\theta = \frac{2a_{kl}^{(m-1)}}{a_{kk}^{(m-1)} - a_{ll}^{(m-1)}}
$$

The value of θ is restricted to

$$
-\frac{\pi}{4} \leqslant \theta \leqslant \frac{\pi}{4}
$$

Thus the orthogonal matrix P_m will appear as

$$
P_m = \begin{array}{c}
\begin{array}{cccccccccccccccc}
 & & & & \begin{matrix} k \\ \text{column} \end{matrix} & & & & & & & \begin{matrix} l \\ \text{column} \end{matrix} & & & &
\end{array} \\
\left[\begin{array}{cccccccccccccccc}
1&&&&&&&&&&&&&&& \\
&1&&&&&&&&&&&&&& \\
&&1&&&&&&&&&&&&& \\
&&&1&&&&&&&&&&&& \\
&&&&\cos\theta&\cdot&\cdot&\cdot&\cdot&\cdot&\cdot&\sin\theta&&&& \\
&&&&&1&&&&&&&&&& \\
&&&&&&1&&&&&&&&& \\
&&&&&&&1&&&&&&&& \\
&&&&&&&&1&&&&&&& \\
&&&&&&&&&1&&&&&& \\
&&&&&&&&&&1&&&&& \\
&&&&-\sin\theta&\cdot&\cdot&\cdot&\cdot&\cdot&\cdot&\cos\theta&&&& \\
&&&&&&&&&&&&1&&& \\
&&&&&&&&&&&&&1&& \\
&&&&&&&&&&&&&&1& \\
&&&&&&&&&&&&&&&1
\end{array}\right]
\end{array}
\begin{array}{l}
\\ \\ \\ \\ \\ k \text{ row} \\ \\ \\ \\ \\ \\ \\ l \text{ row} \\ \\ \\ \\ \\
\end{array}
$$

Since this matrix transformation only makes changes in the k and l rows and columns of the A matrix, the expressions for these new terms can be found from the relationships

$$
\begin{aligned}
a_{ik}^{(m)} &= a_{ki}^{(m)} = a_{ik}^{(m-1)} \cos\theta + a_{il}^{(m-1)} \sin\theta, \qquad \text{for } i \neq k, l \\
a_{il}^{(m)} &= a_{li}^{(m)} = -a_{ik}^{(m-1)} \sin\theta + a_{il}^{(m-1)} \cos\theta, \qquad \text{for } i \neq k, l \\
a_{kk}^{(m)} &= a_{kk}^{(m-1)} \cos^2\theta + 2a_{kl}^{(m-1)} \cos\theta \sin\theta + a_{ll}^{(m-1)} \sin^2\theta \\
a_{ll}^{(m)} &= a_{kk}^{(m-1)} \sin^2\theta - 2a_{kl}^{(m-1)} \cos\theta \sin\theta + a_{ll}^{(m-1)} \cos^2\theta \\
a_{kl}^{(m)} &= (a_{ll}^{(m-1)} - a_{kk}^{(m-1)}) \cos\theta \sin\theta + a_{kl}^{(m-1)}(\cos^2\theta - \sin^2\theta)
\end{aligned}
$$

The values for $\cos\theta$ and $\sin\theta$ can be found from the relationships

$$
\begin{aligned}
x &= 2a_{kl} \\
y &= a_{kk} - a_{ll} \\
z &= \sqrt{x^2 + y^2} \\
\text{sign} &= \frac{x/y}{|(x/y)|}
\end{aligned}
$$

then

$$
\cos\theta = \sqrt{\frac{z+y}{2z}}
$$

$$
\sin\theta = \text{sign}\sqrt{\frac{z-y}{2z}}
$$

If the final rotation matrix is denoted by P_f, then

$$
(P_f \ldots P_2 P_1) A_0 (P_1^T P_2^T \ldots P_f^T) = \text{pure diagonal form}
$$

The eigenvectors of A_0 are therefore the columns of the matrix defined by

$$
R_f = P_1^T P_2^T P_3^T \ldots P_f^T
$$

If the eigenvectors are required, then at each stage of the process the product R_i can be obtained by applying the P_i^T matrix. The starting matrix R_1 would be the identity matrix. Thus the R matrix would be changed only in the k and l columns as

$$
\begin{aligned}
r_{ik}^{(m)} &= r_{ik}^{(m-1)} \cos\theta + r_{il}^{(m-1)} \sin\theta, \qquad i = 1, \ldots, n \\
r_{il}^{(m)} &= -r_{ik}^{(m-1)} \sin\theta + r_{il}^{(m-1)} \cos\theta
\end{aligned}
$$

Before considering other transform methods, an example application of the use of the Jacobi method will be presented.

EXAMPLE 4-2 Suppose that it is desired to find all principal stress values for the state of stress presented in Example 4-1. Suppose further that it is desired to find the eigenvectors associated with these eigenvalues. The need for all eigenvalues might arise if the user were planning to use some failure theory other than the maximum stress theory of failure. Since all eigenvalues are required, the Jacobi method will be used. A BASIC program that solves this problem is given next. The program contains a subroutine that implements the Jacobi method.

```
1000  REM **********************
1010  REM *THIS PROGRAM FINDS   *
1020  REM *THE EIGENVALUES OF   *
1030  REM *A REAL SYMMETRIC     *
1040  REM *MATRIX BY THE JACOBI*
1050  REM *METHOD.              *
1060  REM **********************
1070 :
1080 :
1090  DIM A(4,4),R(4,4)
1100  DIM G$(4),F(4)
1110 N = 3
1120 :
1130 L$ = "-------------------"
1140 M$ = L$ + L$
1150 A(1,1) = 10E06
1160 A(1,2) = 5E06
1170 A(1,3) = 6E06
1180 A(2,1) = A(1,2)
1190 A(2,2) = 20E06
1200 A(2,3) = 4E06
1210 A(3,1) = A(1,3)
1220 A(3,2) = A(2,3)
1230 A(3,3) = 30E06
1240 :
1250  REM *IDENTITY MATRIX R
1260  FOR J = 1 TO 3
1270  FOR I = 1 TO 3
1280 R(I,J) = 0
1290  NEXT I
1300  NEXT J
1310 R(1,1) = 1:R(2,2) = 1:R(3,3)
      = 1
1320 :
1330  PRINT "BEFORE ITERATION"
1340  GOSUB 5000
1350  PRINT : PRINT : PRINT
1360  GOSUB 3000
1370  PRINT "AFTER  ITERATION"
1380  GOSUB 5000
1390  END
1400 :
1410 :
3000  REM **********************
3010  REM * THIS SUBROUTINE     *
3020  REM * APPLIES THE JACOBI  *
3030  REM * METHOD TO           *
3040  REM * DIAGONALIZE A       *
3050  REM * SYMMETRIC MATRIX.   *
3060  REM *                     *
3070  REM * THE RESULT PROVIDES*
3080  REM * ALL EIGENVALUES AND*
3090  REM * EIGENVECTORS OF THE*
3100  REM * MATRIX.             *
3110  REM *                     *
3120  REM *                     *
3130  REM * THE PROCESS STOPS   *
3140  REM * WHEN THE LARGEST    *
3150  REM * OFF-DIAGONAL ELE-   *
3160  REM * MENT IS SMALLER     *
3170  REM * THAN 0.0001 TIMES   *
3180  REM * THE AVG DIAGONAL    *
3190  REM * TERM.               *
3200  REM *                     *
3210  REM *                     *
3220  REM *  PARAMETERS:        *
3230  REM *                     *
3240  REM *   N     - NUMBER OF*
3250  REM *           ROWS AND *
3260  REM *           COLUMNS  *
3270  REM *           IN MATRIX*
3280  REM *                     *
3290  REM *   A     - ORIGINAL *
3300  REM *           MATRIX   *
3310  REM *           DIMENSION*
3320  REM *           (N+1,N+1)*
```

```
3330  REM *                      *
3340  REM *                      *
3350  REM *   R      - WORKING   *
3360  REM *            ARRAY OF  *
3370  REM *            SAME SIZE*
3380  REM *            AS "A."   *
3390  REM *            STARTS AS*
3400  REM *            IDENTITY  *
3410  REM *            MATRIX.   *
3420  REM *                      *
3430  REM *                      *
3440  REM * ON COMPLETION THE    *
3450  REM * EIGENVALUES ARE THE*
3460  REM * DIAGONAL ELEMENTS    *
3470  REM * OF MATRIX "A" AND    *
3480  REM * THE EIGENVECTORS     *
3490  REM * ARE THE COLUMNS OF  *
3500  REM * THE "R" MATRIX.      *
3510  REM *                      *
3520  REM ***********************
3530 :
3540 :
3550  REM **FIND THE LARGEST ***
3560  REM **OFF-DIAGONAL TERM***
3570 :
3580 AMAX =  ABS (A(1,2)):RM = 1:
      CM = 2
3590  FOR R = 1 TO N - 1
3600  FOR C = R + 1 TO N
3610  IF  ABS (A(R,C))< = AMAX GOTO
      3630
3620 RM = R:CM = C:AMAX =  ABS (A
      (R,C))
3630  NEXT C
3640  NEXT R
3650 :
3660 :
3670  REM ***FIND THE AVERAGE***
3680  REM ***MAGNITUDE OF THE***
3690  REM ***DIAGONAL TERMS  ***
3700 :
3710 SUM = 0
3720  FOR I = 1 TO N
3730 SUM = SUM +  ABS (A(I,I))
3740  NEXT I
3750 AVG = SUM / N
3760 :
3770 :
3780 :
3790  IF AMAX < 1E - 04 * AVG THEN
      GOTO 4280
3800 :
3810 :
3820  REM ***TRANSFORM THE    ***
3830  REM ***MATRIX TO ZERO   ***
3840  REM ***THE LARGEST OFF-***
3850  REM ***DIAGONAL ELEMENT***
3860 :
3870  REM **SOLVE FOR THETA**
3880 :
3890 X = 2 * A(RM,CM)
3900 Y = A(RM,RM) - A(CM,CM)
3910 SIGN = (X / Y) /  ABS (X / Y
      )
3920 Z =  SQR (X * X + Y * Y)
3930 C =  SQR ((Z + Y) / (2 * Z))

3940 S = SIGN *  SQR ((Z - Y) / (
      2 * Z))
3950 :
3960 :
3970  REM ***CALCULATE NEW    ***
3980  REM ***MATRIX TERMS     ***
3990 :
4000  FOR I = 1 TO N
4010  IF I = RM GOTO 4080
4020  IF I = CM GOTO 4080
4030 B = A(I,RM) * C + A(I,CM) *
      S
4040 A(I,CM) =  - A(I,RM) * S + A
      (I,CM) * C
4050 A(CM,I) = A(I,CM)
4060 A(I,RM) = B
4070 A(RM,I) = B
4080  NEXT I
4090 B = A(RM,RM) * C * C + 2 * A
      (RM,CM) * C * S + A(CM,CM) *
      S * S
4100 Z = (A(CM,CM) - A(RM,RM)) *
      C * S + A(RM,CM) * (C * C -
      S * S)
4110 A(CM,CM) = A(RM,RM) * S * S -
      2 * A(RM,CM) * C * S + A(CM,
      CM) * C * C
4120 A(RM,RM) = B
4130 A(RM,CM) = Z:A(CM,RM) = Z
4140 :
4150 :
4160  REM ***UPDATE THE  R    ***
4170  REM ***MATRIX OF EIGEN-***
4180  REM ***VECTORS          ***
4190 :
4200  FOR I = 1 TO N
4210 B = R(I,RM) * C + R(I,CM) *
      S
4220 R(I,CM) =  - R(I,RM) * S + R
      (I,CM) * C
4230 R(I,RM) = B
4240  NEXT I
4250 :
4260 :
4270  GOTO 3000
4280  RETURN
4290 :
```

```
4300 :
5000  REM *********************
5010  REM * THIS SUBROUTINE    *
5020  REM * PRINTS THE RESULTS *
5030  REM *********************
5040 :
5050  REM *** PRINT THE TWO  ***
5060  REM *** MATRICES, A & R***
5070 :
5080  PRINT M$
5090  PRINT
5100  PRINT "THE A MATRIX IS"
5110  PRINT
5120  FOR J = 1 TO 3
5130  FOR I = 1 TO 3
5140 F(I) = A(J,I)
5150  NEXT I
5160  GOSUB 6000
5170  PRINT G$(1); TAB( 15);G$(2)
     ; TAB( 29);G$(3)
5180  NEXT J
5190  PRINT M$
5200  PRINT
5210  PRINT "THE R MATRIX IS"
5220  PRINT
5230  FOR J = 1 TO N
5240  FOR I = 1 TO 3
5250 F(I) = R(J,I)
5260  NEXT I
5270  GOSUB 6000
5280  PRINT G$(1); TAB( 14);G$(2)
     ; TAB( 29);G$(3)
5290  NEXT J
5300  PRINT M$
5310  RETURN
5320 :
5330 :
6000  REM ********************
6010  REM * THIS SUBROUTINE   *
6020  REM * SORTS SCIENTIFIC  *
6030  REM * NOTATION FOR PRINT*
6040  REM * CLARITY.          *
6050  REM ********************
6060 :
6070 :
6080  FOR I = 1 TO N
6090 D$ =  STR$ (F(I))
6100  IF  LEN (D$) < = 10 THEN  GOTO
     6140
6110  IF  ABS (F(I)) > 0.01 THEN
      GOTO 6140
6120 G$(I) =  LEFT$ (D$,6) +  RIGHT$
     (D$,4)
6130  GOTO 6150
6140 G$(I) =  LEFT$ (D$,10)
6150  NEXT I
6160  RETURN
```

The output from this program is as follows.

```
BEFORE ITERATION
-------------------------------------

THE A MATRIX IS

10000000       5000000        6000000
5000000        20000000       4000000
6000000        4000000        30000000
-------------------------------------


THE R MATRIX IS

1              0              0
0              1              0
0              0              1
-------------------------------------


AFTER  ITERATION
-------------------------------------

THE A MATRIX IS

33709178.5     -.78963872     555.004529
-.78963872     19149061.2     1.88710195
555.004529     1.88710195     7141760.31
-------------------------------------
```

```
THE R MATRIX IS

-.30017268    .196732957     .933376937
-.36643156    .879640306     -.30325045
-.88069533    -.43304627     -.19195482
--------------------------------------
```

The eigenvalues appear in the output as diagonal elements of the A matrix, and the eigenvectors appear as columns of the R matrix. This program required less than 10 seconds to run on an Apple II computer. An Apple II computer with 48K of memory could handle this program with a matrix having more than 50 rows and columns. As mentioned in Example 4-1, larger matrices could be handled if the user could store the A matrix in a more streamlined form. The capacity of the program could also be expanded if the eigenvectors were not required. The program contains two special subroutines to permit simple formatting of the output for the sake of readability.

Given's Method for Symmetric Matrices

Given's method is based on similarity transformations of the same type as used for Jacobi's method; however, the scheme is arranged so that once zeros are created in an element, they are retained in later transformations. For this reason the method requires only a fixed finite number of transformations. Thus, the method is more efficient in computer time when compared with the Jacobi method. The only disadvantage to the method is that it produces a tridiagonal form for a symmetric matrix rather than a diagonal form. We shall see later that the tridiagonal form is quite useful and is well worth achieving.

When applied to an $n \times n$ matrix, the Given's method uses $n - 2$ major steps. Each of these major steps may require several transformations depending on the number of zeros that must be introduced into a given column or row. In the kth step, zeros are introduced into the nontridiagonal elements of the kth row and kth column without destroying the zeros produced during the previous $k - 1$ steps. Thus, as it begins the kth step, the transformed matrix is tridiagonal as far as its first $k - 1$ rows and columns are concerned. The pattern of transformed matrices for a symmetric 5×5 system would be

$A_0 =$ 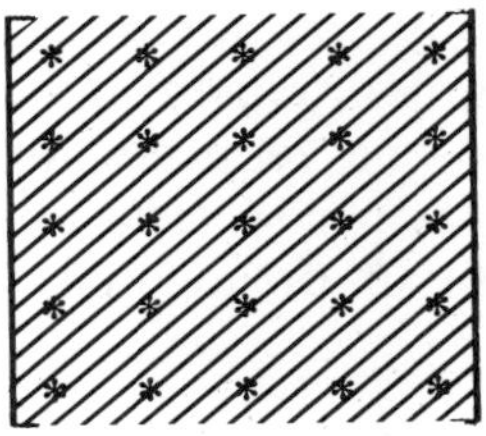 (the starting matrix)

$$A_1 = \begin{bmatrix} * & * & 0 & 0 & 0 \\ * & * & * & * & * \\ 0 & * & * & * & * \\ 0 & * & * & * & * \\ 0 & * & * & * & * \end{bmatrix}$$ (after the first major step consisting of three separate transformations)

$$A_2 = \begin{bmatrix} * & * & 0 & 0 & 0 \\ * & * & * & 0 & 0 \\ 0 & * & * & * & * \\ 0 & 0 & * & * & * \\ 0 & 0 & * & * & * \end{bmatrix}$$ (after the second major step consisting of two separate transformations)

$$A_3 = \begin{bmatrix} * & * & 0 & 0 & 0 \\ * & * & * & 0 & 0 \\ 0 & * & * & * & 0 \\ 0 & 0 & * & * & * \\ 0 & 0 & 0 & * & * \end{bmatrix}$$ (after the third major step consisting of one transformation; the matrix is now in tridiagonal form)

In each major step, only elements a_{ij} that are in the lower right shaded area are changed. Thus, in the kth step we are concerned only with the matrix of order $(n - k + 1)$ in the bottom right corner of the previous matrix. Clearly, as the stages proceed, the number of transformations per stage decreases. The total number of transformations required to reach the tridiagonal form will be $(n^2 - 3n + 2)/2$.

Based on our observations of Given's method, it seems reasonable to desire a method that will reduce a whole row and column of non-tridiagonal terms to zero at the same time. A method that accomplishes this has been presented by Householder.

Householder's Method for Symmetric Matrices

Because it transforms whole rows and columns of nontridiagonal elements to zero simultaneously, the method of Householder achieves a tridiagonal form in about half as many computations. Even though Householder's method uses a more complex transformation, it is usually faster than Given's method owing to the reduced number of transformations necessary. This efficiency is particularly obvious for large matrices. Although it utilizes Hermitian orthogonal transformation matrices rather than plane rotations, the tridiagonal form generated by Householder's method will have the same eigenvalues as the tridiagonal form generated by Given's method. The transformations involved in the $n - 2$ major steps of this method are

$$A_k = P_k A_{k-1} P_k^T, \qquad k = 1, \ldots, n - 2$$

where $A_0 = A$. The transformation matrices will each be of the form

$$P_k = I - \frac{u_k u_k^T}{2K_k^2}$$

where

$$\begin{aligned} u_{ik} &= 0, & i &= 1, 2, \ldots, k \\ u_{ik} &= a_{ki}, & i &= k + 2, \ldots, n \\ u_{k+1,k} &= a_{k,k+1} + S_k \end{aligned}$$

In these elements

$$S_k = \left[\sum_{i=k+1}^{n} a_{ki}^2 \right]^{1/2}$$

and
$$2K_k^2 = S_k^2 \mp a_{k,k+1} S_k$$

The sign choice in these equations is taken to be the same as that for $a_{k,k+1}$. In this way the value of $u_{k+1,k}$ is maximized. It should be noted that Given's and Householder's methods can be applied to nonsymmetric matrices. The result will not be a tridiagonal form but will be a special form known as the Hessenberg form. The Hessenberg form is a matrix that is almost triangular.

If the matrix takes the form

$$\begin{bmatrix} * & * & 0 & 0 & 0 & 0 \\ * & * & * & 0 & 0 & 0 \\ * & * & * & * & 0 & 0 \\ * & * & * & * & * & 0 \\ * & * & * & * & * & * \\ * & * & * & * & * & * \end{bmatrix}$$

it is called lower Hessenberg. If the matrix takes the form

$$\begin{bmatrix} * & * & * & * & * & * \\ * & * & * & * & * & * \\ 0 & * & * & * & * & * \\ 0 & 0 & * & * & * & * \\ 0 & 0 & 0 & * & * & * \\ 0 & 0 & 0 & 0 & * & * \end{bmatrix}$$

it is called upper Hessenberg.

These two special forms are extremely useful as an intermediate step to finding all eigenvectors of a general matrix. The methods of Given and Householder are not the only ways of converting a general matrix to Hessenberg form. In Section 4.5 we will look at a special direct method for doing this for a general matrix, but first let us complete the topic of finding eigenvalues of a symmetric tridiagonal matrix.

4.4 FINDING THE EIGENVALUES OF A SYMMETRIC TRIDIAGONAL MATRIX

Once Given's or Householder's method has been applied to a symmetric matrix and a tridiagonal form has been achieved, it next becomes necessary to find the eigenvalues. To see the utility of the tridiagonal form, the eigenvalue problem will be written in the form

$$\det (A - \lambda I) = 0$$

where A is a symmetric tridiagonal matrix. This system will look as follows:

$$\begin{bmatrix} a_1 - \lambda & b_2 & & & 0 \\ b_2 & a_2 - \lambda & & & \\ & \cdot & & & \\ & & \cdot & & \\ & & & \cdot & \\ & & & & b_n \\ & 0 & & b_n & a_n - \lambda \end{bmatrix} = 0$$

The expansion of a general $n \times n$ determinant reduces to a system of n subdeterminants each $(n - 1) \times (n - 1)$. Each n subdeterminant will give rise to $n - 1$ subdeterminants each $(n - 2) \times (n - 2)$. The preceding tridiagonal form is fortunate since only two of the subdeterminants are nonzero at each step. Thus the general determinant can be expressed as a sequence of polynomials:

$$f_m(\lambda) = (a_m - \lambda) f_{m-1}(\lambda) - b_m^2 f_{m-2}(\lambda)$$

If one lets

$$f_0(\lambda) = 1$$

and

$$f_1(\lambda) = a_1 - \lambda$$

then for $r = 2, \ldots, n$ we get a sequence of polynomials known as the Sturm sequence. One property of this special type of sequence is that the roots of the $f_j(\lambda)$ polynomial separate the roots of the $f_{j+1}(\lambda)$ polynomial. Thus, for $f_1(\lambda) = a_1 - \lambda = 0$, one can predict that the value $\lambda = a_1$ separates the two roots of $f_2(\lambda) = (a_2 - \lambda)(a_1 - \lambda) - b_2^2 = 0$. This information about the location of the roots of $f_2(\lambda)$ makes the iterative solution of this polynomial an easy process. Indeed the technique of binary search can be utilized if one knows bounds for the roots of the polynomial. As the process proceeds, the sequence leads to the solution of the final polynomial $f_n(\lambda) = 0$, which will give the n eigenvalues. This process can be illustrated by placing the roots of the sequence of polynomials as follows:

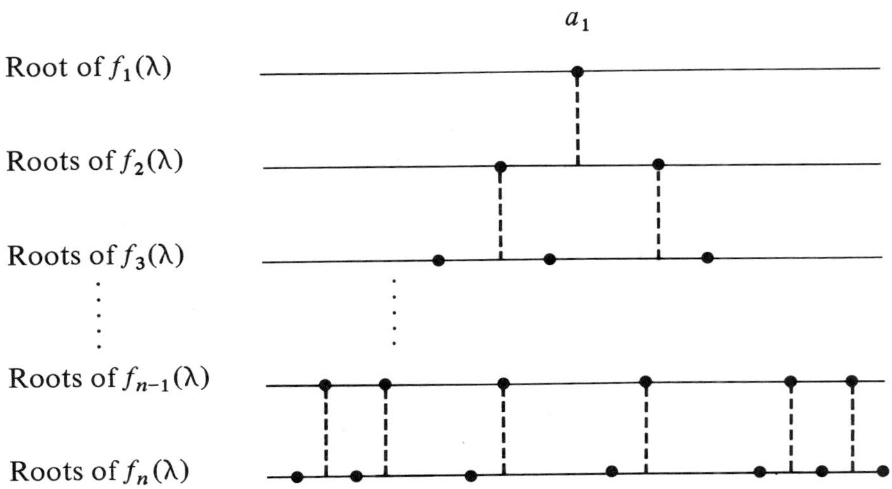

A further property of the Sturm sequence is that, for any value of b such that $f_n(b) \neq 0$, the number of eigenvalues of A that are greater than b is equal to the number of sign changes in the sequence

$$1, -f_1(b), f_2(b), \ldots, (-1)^n f_n(b)$$

If this integer number of sign changes is called $V(b)$, the number of eigenvalues between two real points b and c is $V(b) - V(c)$.

4.5 DIRECT REDUCTION OF A MATRIX TO HESSENBERG FORM

In changing a general matrix to Hessenberg form, we seek a transformation matrix N such that

$$N^{-1}AN = H$$

where A is the original matrix and H is in the upper Hessenberg form. If this expression is premultiplied by the matrix N, we see

$$AN = NH$$

If the N matrix is selected to be a special lower triangular form, the matrix equation will become

$$\overset{A}{\begin{bmatrix} * & * & * & * & * \\ * & * & * & * & * \\ * & * & * & * & * \\ * & * & * & * & * \\ * & * & * & * & * \end{bmatrix}} \overset{N}{\begin{bmatrix} 1 & & & & \\ 0 & 1 & & & \\ 0 & * & 1 & & \\ 0 & * & * & 1 & \\ 0 & * & * & * & 1 \end{bmatrix}}$$

$$= \overset{N}{\begin{bmatrix} 1 & & & & \\ 0 & 1 & & & \\ 0 & * & 1 & & \\ 0 & * & * & 1 & \\ 0 & * & * & * & 1 \end{bmatrix}} \overset{H}{\begin{bmatrix} * & * & * & * & * \\ * & * & * & * & * \\ & * & * & * & * \\ & & * & * & * \\ & & & * & * \end{bmatrix}}$$

A careful inspection of these matrices reveals that there are n^2 terms to be determined. These terms are the $(n^2 + 3n - 2)/2$ terms of the H matrix, and the remaining $(n^2 - 3n + 2)/2$ are terms of the N

matrix. These unknowns can be found by equating elements on both sides of the equation. This process can be handled by any of the elimination methods presented in Chapter 3. A straightforward technique to accomplish this has been suggested by Wilkinson [10] and is as follows:

Step 1

For each value of i from 1 to r, compute

$$h_{ir} = a_{ir} + \sum_{k=r+1}^{n} a_{ik} n_{kr} - \sum_{k=2}^{i-1} n_{ik} h_{kr}$$

In order to save space, the value of h_{ir} may be overwritten on a_{ir}.

Step 2

For each value of i from $r + 1$ to n, compute

$$\text{temp}_i = a_{ir} + \sum_{k=r+1}^{n} a_{ik} n_{kr} - \sum_{k=2}^{r} n_{ik} h_{kr}$$

As successive values of temp_i are computed, a record should be kept of which value has the largest magnitude and the i location of this value, i_{large}.

Step 3

Interchange rows $r + 1$ and i_{large} (including n_{ij} values and temp_i values), and then interchange columns $r + 1$ and i_{large}. The effect of this interchange is to bring the largest possible pivot element to the position $h_{r+1,r}$ for use in the next step. The rationale for this process is the same as that used for simultaneous algebraic equations in Chapter 3, that is, to ensure maximum accuracy and to be sure that a zero pivot element is not used. It can be shown that the eigenvalues of a matrix remain unchanged when two rows are interchanged only if the corresponding columns are also interchanged.

Step 4

Take the current temp_{r+1} element and overwrite it on $a_{r+1,r}$. This is the pivotal element.

Step 5

For each value of i from $r + 2$ to n, compute

$$n_{i,r+1} = \frac{\text{temp}_i}{h_{r+1,r}}$$

and overwrite on a_{ir}. Thus we see that the two matrices N and H can be stored in the space required for the original matrix A if the user is willing to shift the N matrix to the left one column and remember that the first column of N is nearly all zeros and that the diagonal of N is all ones.

This elimination scheme is an efficient method for converting a general matrix to Hessenberg form. In Example 4-3 we will consider a subroutine that implements this procedure, but first let us consider how the Hessenberg form can be used to achieve the eigenvalues themselves.

4.6 OTHER METHODS OF EIGENVALUE CALCULATION

In this section, two significant methods for finding eigenvalues will be discussed. The methods were both developed within the past 20 years and represent the most efficient methods presently available when all eigenvalues of a general real or complex matrix are desired. Both methods use transformations to produce a sequence of similar matrices that converge to a block triangular form:

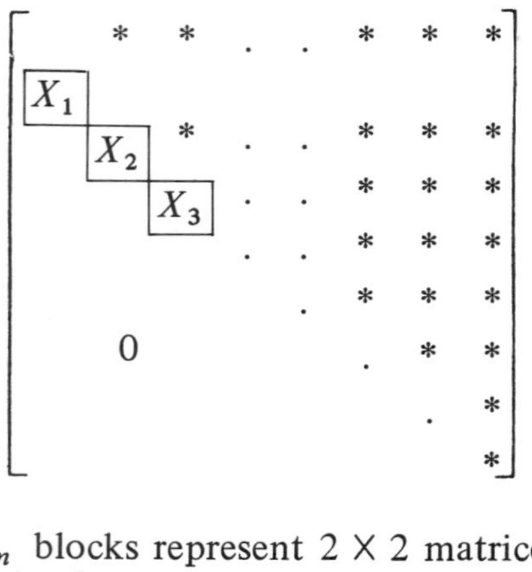

In this form the X_m blocks represent 2×2 matrices on the diagonal. The eigenvalues of the X_m blocks are the eigenvalues of the original $n \times n$ matrix. This form is convenient because the second-order determinant of the X_m blocks allows complex eigenvalues to be found without the need for complex terms in the final matrix. If all eigenvalues of the original matrix are pure real, the final form will be pure triangular with the eigenvalues on the diagonal.

The LR Method

The *LR* method was first developed by Rutishauser in 1958. In this method a matrix A is decomposed into

$$A = LR$$

where L is unit lower triangular and R is upper triangular. Using the similarity transformation $L^{-1}AL$, we see

$$A_2 = L^{-1}AL = L^{-1}(LR)L = RL$$

Thus

$$A_{m-1} = L_{m-1}R_{m-1}$$

and

$$A_m = R_{m-1}L_{m-1}$$

This process is repeated until L_s tends to the identity matrix I and R_s tends to A_s, which tends to the block diagonal form. Although this method is easy to apply, it may become unstable. For this reason the method discussed next is often preferred.

The QR Method

The *QR* method was first presented by Francis in 1961 [4]. The *QR* algorithm is defined by the relationship

$$A_m = Q_m R_m$$

In this expression, Q_m is an orthogonal matrix and R_m is an upper triangular matrix. As the method progresses

$$A_{m+1} = Q_m^T A_m Q_m = Q_m^T Q_m R_m Q_m = R_m Q_m$$

In the limit, the sequence of A matrices becomes the block diagonal form. This method is more difficult to implement and consumes more computer time than the *LR* method. Nevertheless, its stable numerical performance caused by the use of orthogonal transformation matrices causes it to be regarded as the best general solution method.

As it was originally developed by Francis, acceleration of the convergence of the QR algorithm has been shown to be possible if shifts of the origin as described by the equation

$$A_m - k_m I = Q_m R_m$$

are used. In this relationship, Q_m is the orthogonal transformation matrix, R_m is upper triangular, and k_m is the shift of origin. When the original matrix A_1 is of upper Hessenberg form, it can be shown that the volume of work to reach final convergence is far less. Since a general real coefficient matrix A can have complex eigenvalues, the QR procedure can become difficult for small computers since most are unable to perform complex manipulations. This difficulty can be overcome by performing two steps with shifts k_m and k_{m+1} that are either pure real values or are complex conjugate values. Francis has described an excellent method that allows the user to implement this procedure in order to avoid the use of complex arithmetic. An example that makes use of this technique for the QR algorithm will now be presented.

EXAMPLE 4-3 Suppose that it is desired to find all eigenvalues of the general 6 X 6 matrix

$$\begin{bmatrix} 2.3 & 4.3 & 5.6 & 3.2 & 1.4 & 2.2 \\ 1.4 & 2.4 & 5.7 & 8.4 & 3.4 & 5.2 \\ 2.5 & 6.5 & 4.2 & 7.1 & 4.7 & 9.3 \\ 3.8 & 5.7 & 2.9 & 1.6 & 2.5 & 7.9 \\ 2.4 & 5.4 & 3.7 & 6.2 & 3.9 & 1.8 \\ 1.8 & 1.7 & 3.9 & 4.6 & 5.7 & 5.9 \end{bmatrix}$$

The eigenvalues will be found by a two-step process in which the direct reduction method of Section 4.5 will be used to reduce the matrix to upper Hessenberg form, and then the QR algorithm will be used to find all eigenvalues.

The following BASIC program implements the solution. The subroutine starting at line 3000 performs the reduction to Hessenberg form, and the subroutine starting at line 5000 performs the QR algorithm. This subroutine was adapted from an ALGOL algorithm presented by Martin, Peters, and Wilkinson [7]. Since the subroutine uses double QR steps, complex eigenvalues can be found without the need for complex algebraic manipulations.

```
1000  REM *********************
1010  REM *THIS PROGRAM FINDS  *
1020  REM *THE EIGENVALUES OF  *
1030  REM *A GENERAL MATRIX    *
1040  REM *BY REDUCTION TO     *
1050  REM *HESSENBERG FORM AND *
1060  REM *THEN APPLYING THE   *
1070  REM *QR ALGORITHM        *
1080  REM *********************
1090 :
1100 :
1110  DIM A(7,7),G$(7),F(7)
1120  DIM RR(7),RI(7):N = 6
1130 :
1140  DATA  2.3,4.3,5.6,3.2,1.4,2
     .2
1150  DATA  1.4,2.4,5.7,8.4,3.4,5
     .2
1160  DATA  2.5,6.5,4.2,7.1,4.7,9
     .3
1170  DATA  3.8,5.7,2.9,1.6,2.5,7
     .9
1180  DATA  2.4,5.4,3.7,6.2,3.9,1
     .8
1190  DATA  1.8,1.7,3.9,4.6,5.7,5
     .9
1200  DATA  1,0,0,0,0,0
1210 :
1220  FOR I = 1 TO 6
1230  READ A(I,1),A(I,2),A(I,3),A
     (I,4),A(I,5),A(I,6)
1240  NEXT I
1250  READ A(1,0),A(2,0),A(3,0),A
     (4,0),A(5,0),A(6,0)
1260 :
1270  PRINT : PRINT : PRINT
1280  PRINT "THE ORIGINAL MATRIX
     IS"
1290  GOSUB 7000
1300  GOSUB 3000
1310  PRINT : PRINT : PRINT
1320  PRINT "IN UPPER HESSENBERG
     FORM": GOSUB 7000
1330  GOSUB 5000
1340  PRINT : PRINT : PRINT
1350  PRINT "THE EIGENVALUES ARE"

1360 M$ = "----------"
1370 M$ = M$ + M$ + M$
1380  PRINT M$
1390  PRINT "  REAL PART";"    IMA
     GINARY PART"
1400  PRINT M$
1410  FOR I = 1 TO N
1420  PRINT RR(I),RI(I)
1430  NEXT I
1440  PRINT M$
1450  END
1460 :
1470 :
3000  REM *********************
3010  REM * THIS SUBROUTINE    *
3020  REM * REDUCES THE GENERAL*
3030  REM * MATRIX TO HESSEN-  *
3040  REM * BERG FORM BY DIRECT*
3050  REM * REDUCTION USING    *
3060  REM * INTERCHANGES TO    *
3070  REM * MINIMIZE ROUNDOFF  *
3080  REM * ERROR.             *
3090  REM *                    *
3100  REM *   PARAMETERS:      *
3110  REM *                    *
3120  REM *    A - THE ORIGINAL*
3130  REM *        MATRIX OF   *
3140  REM *        DIMENSION   *
3150  REM *        (N+1,N+1)   *
3160  REM *        ON RETURN IT*
3170  REM *        CONTAINS THE*
3180  REM *        UPPER       *
3190  REM *        HESSENBERG  *
3200  REM *        FORM.       *
3210  REM *                    *
3220  REM *    N - ORDER  OF   *
3230  REM *        THE MATRIX. *
3240  REM *                    *
3250  REM *********************
3260 :
3270 :
3280 :
3290  FOR R = 1 TO N
3300  FOR I = 1 TO R
3310 SUM = 0
3320  IF R + 1 > N THEN  GOTO 336
     0
3330  FOR K = R + 1 TO N
3340 SUM = SUM + A(I,K) * A(K,R -
     1)
3350  NEXT K
3360  IF 2 > I - 1 THEN  GOTO 340
     0
3370  FOR K = 2 TO I - 1
3380 SUM = SUM - A(I,K - 1) * A(K
     ,R)
3390  NEXT K
3400 A(I,R) = SUM + A(I,R)
3410  NEXT I
3420 :
3430 :
3440  IF R + 1 > N THEN  GOTO 357
     0
3450  FOR I = R + 1 TO N
3460 SUM = 0
3470  IF R + 1 > N THEN  GOTO 351
     0
3480  FOR K = R + 1 TO N
3490 SUM = SUM + A(I,K) * A(K,R -
     1)
3500  NEXT K
```

```
3510  IF 2 > R THEN  GOTO 3550
3520  FOR K = 2 TO R
3530 SUM = SUM - A(I,K - 1) * A(K
     ,R)
3540  NEXT K
3550 A(0,I) = SUM + A(I,R)
3560  NEXT I
3570 :
3580 :
3590  REM ***FIND THE LARGEST***
3600  REM ***A(0,I) ELEMENT  ***
3610  REM ***TO USE FOR PIVOT***
3620 :
3630  IF R + 1 > N THEN  GOTO 395
     0
3640 AMAX =  ABS (A(0,R + 1)):RM =
     R + 1
3650  FOR I = R + 1 TO N
3660  IF  ABS (A(0,I)) < AMAX THEN
      GOTO 3680
3670 AMAX =  ABS (A(0,I)):RM = I
3680  NEXT I
3690 :
3700 :
3710  REM ***WHEN RM=R+1 SKIP***
3720  REM ***THE INTERCHANGE ***
3730 :
3740  IF RM = R + 1 THEN  GOTO 39
     50
3750 :
3760 :
3770  REM ***INTERCHANGE THE ***
3780  REM ***COLUMNS R+1 & RM***
3790 :
3800  FOR I = 0 TO N
3810 TEMP = A(I,R + 1)
3820 A(I,R + 1) = A(I,RM)
3830 A(I,RM) = TEMP
3840  NEXT I
3850 :
3860 :
3870  REM ***INTERCHANGE THE ***
3880  REM ***ROWS R+1 AND RM ***
3890  FOR I = 1 TO N
3900 TEMP = A(R + 1,I)
3910 A(R + 1,I) = A(RM,I)
3920 A(RM,I) = TEMP
3930  NEXT I
3940 :
3950 A(R + 1,R) = A(0,R + 1)
3960 :
3970  IF R + 2 > N THEN  GOTO 401
     0
3980  FOR I = R + 2 TO N
3990 A(I,R) = A(0,I) / A(R + 1,R)

4000  NEXT I
4010  NEXT R
4020  FOR J = 1 TO N - 2
4030  FOR I = J + 2 TO N
4040 A(I,J) = 0
4050  NEXT I
4060  NEXT J
4070  RETURN
4080 :
4090 :
5000  REM *********************
5010  REM * THIS SUBROUTINE    *
5020  REM * FINDS  THE  EIGEN- *
5030  REM * VALUES OF A REAL,  *
5040  REM * UPPER  HESSENBERG  *
5050  REM * ALGORITHM USING    *
5060  REM * SHIFTS OF ORIGIN TO*
5070  REM * ACCELERATE CONVER- *
5080  REM * GENCE AS SUGGESTED *
5090  REM * BY FRANCIS.        *
5100  REM *                    *
5110  REM *   PARAMETERS:      *
5120  REM *                    *
5130  REM *     A - THE STARTING*
5140  REM *         ARRAY OF UP-*
5150  REM *         PER HESSEN- *
5160  REM *         BERG FORM.  *
5170  REM *         DIMENSIONED *
5180  REM *         (N+1,N+1).  *
5190  REM *                    *
5200  REM *     N - ORDER  OF   *
5210  REM *         THE MATRIX. *
5220  REM *                    *
5230  REM *    RI - VECTOR OF   *
5240  REM *         IMAGINARY   *
5250  REM *         PARTS OF THE*
5260  REM *         EIGENVALUES *
5270  REM *         DIMENSIONED *
5280  REM *         (N+1).      *
5290  REM *                    *
5300  REM *    RR - VECTOR OF   *
5310  REM *         REAL PARTS  *
5320  REM *         OF THE      *
5330  REM *         EIGENVALUES *
5340  REM *         DIMENSIONED *
5350  REM *         (N+1).      *
5360  REM *                    *
5370  REM * THE SUBROUTINE     *
5380  REM * TERMINATES IF 30   *
5390  REM * ITERATIONS FAIL TO *
5400  REM * PROVIDE CONVERGENCE*
5410  REM *********************
5420 :
5430 :
5440 NN = N
5450  IF NN = 0 THEN  GOTO 6720
5460 IT = 0:NA = NN - 1:MACH = 10
     E - 06
5470 MACH = 1E - 11
5480 :
```

```
5490  REM ***LOOK FOR A SMALL
5500  REM ***SUBDIAGONAL ELEMENT
5510  REM ***STARTING WITH THE
5520  REM ***BOTTOM OF CURRENT
5530  REM ***MATRIX AND WORKING
5540  REM ***UPWARD.
5550 :
5560  FOR L = NN TO 2 STEP  - 1
5570  IF  ABS (A(L,L - 1)) <  = M
    ACH * ( ABS (A(L - 1,L - 1))
     +  ABS (A(L,L))) THEN  GOTO
    5600
5580  NEXT L
5590 L = 1
5600 X = A(NN,NN)
5610  IF L = NN THEN  GOTO 6480
5620 Y = A(NA,NA):R = A(NN,NA) *
    A(NA,NN)
5630  IF NA = L THEN  GOTO 6530
5640  IF IT = 30 THEN  GOTO 6710
5650 :
5660 :
5670  REM ***FORM THE SHIFT
5680 :
5690  IF IT = 10 THEN  GOTO 5730
5700  IF IT = 20 THEN  GOTO 5730
5710 S = X + Y:Y = X * Y - R
5720  GOTO 5750
5730 Y =  ABS (A(NN,NA)) +  ABS (
    A(NA,NN - 2))
5740 S = 1.5 * Y:Y = Y ^ 2
5750 IT = IT + 1
5760 :
5770 :
5780  REM ***LOOK FOR TWO CON-
5790  REM ***SECUTIVE SMALL SUB-
5800  REM ***DIAGONAL ELEMENTS.
5810 :
5820  FOR M = NN - 2 TO L STEP  -
    1
5830 X = A(M,M):R = A(M + 1,M)
5840 Z = A(M + 1,M + 1)
5850 P = X * (X - S) + Y + R * A(
    M,M + 1)
5860 Q = R * (X + Z - S):R = R *
    A(M + 2,M + 1)
5870 W =  ABS (P) +  ABS (Q) +  ABS
    (R)
5880 P = P / W:Q = Q / W:R = R /
    W
5890  IF M = L THEN  GOTO 5920
5900  IF  ABS (A(M,M - 1)) * ( ABS
    (Q) +  ABS (R)) <  = MACH *
     ABS (P) * ( ABS (A(M - 1,M -
    1)) +  ABS (X) + ABS (Z)) THEN
     GOTO 5920
5910  NEXT M
5920  FOR I = M + 2 TO NN
5930 A(I,I - 2) = 0
5940  NEXT I
5950  FOR I = M + 3 TO NN
5960 A(I,I - 3) = 0
5970  NEXT I
5980 :
5990 :
6000  REM  ***DO DOUBLE QR STEP
6010  REM  ***ON ROWS L TO NN AND

6020  REM  ***COLUMNS M TO NN
6030 :
6040  FOR K = M TO NA
6050  IF K = M THEN  GOTO 6120
6060 P = A(K,K - 1):Q = A(K + 1,K
     - 1)
6070 R = A(K + 2,K - 1)
6080  IF NA = K THEN R = 0
6090 X =  ABS (P) +  ABS (Q) + ABS
    (R)
6100  IF X = 0 THEN  GOTO 6440
6110 P = P / X:Q = Q / X:R = R /
    X
6120 S =  SQR (P * P + Q * Q + R *
    R)
6130  IF P < 0 THEN S =  - S
6140  IF K <  > M THEN A(K,K - 1)
     =  - S * X
6150  IF L <  > M THEN A(K,K - 1)
     =  - A(K,K - 1)
6160 P = P + S:X = P / S:Y = Q /
    S:Z = R / S
6170 Q = Q / P:R = R / P
6180 :
6190 :
6200  REM ***ROW MODIFICATION
6210 :
6220  FOR J = K TO NN
6230 P = A(K,J) + Q * A(K + 1,J)
6240  IF NA = K THEN  GOTO 6270
6250 P = P + R * A(K + 2,J)
6260 A(K + 2,J) = A(K + 2,J) - P *
    Z
6270 A(K + 1,J) = A(K + 1,J) - P *
    Y
6280 A(K,J) = A(K,J) - P * X
6290  NEXT J
6300 J = NN
6310  IF K + 3 < NN THEN J = K +
    3
6320 :
6330 :
6340  REM ***COLUMN MODIFICATION
6350 :
6360  FOR I = L TO J
6370 P = X * A(I,K) + Y * A(I,K +
    1)
6380  IF NA = K THEN  GOTO 6410
```

```
6390 P = P + Z * A(I,K + 2)
6400 A(I,K + 2) = A(I,K + 2) - P *
     R
6410 A(I,K + 1) = A(I,K + 1) - P *
     Q
6420 A(I,K) = A(I,K) - P
6430  NEXT I
6440  NEXT K
6450  GOTO 5560
6460 :
6470 :
6480  REM ***ONE ROOT FOUND
6490 RR(NN) = X:RI(NN) = 0:NN = N
     A
6500  GOTO 5450
6510 :
6520 :
6530  REM ***TWO ROOTS FOUND
6540 P = (Y - X) / 2:Q = P * P +
     R:Y =  SQR ( ABS (Q))
6550  IF Q < 0 THEN  GOTO 6650
6560 :
6570  REM ***REAL PAIR
6580  IF P < 0 THEN Y =  - Y
6590 Y = P + Y
6600 RR(NA) = X + Y:RR(NN) = X -
     R / Y
6610 RI(NA) = 0:RI(NN) = 0
6620  GOTO 6690
6630 :
6640 :
6650  REM ***COMPLEX PAIR
6660 :
6670 RR(NA) = X + P:RR(NN) = X +
     P
6680 RI(NA) = Y:RI(NN) =  - Y
6690 NN = NN - 2
6700  GOTO 5450
6710  PRINT "NO CONVERGENCE AFTER
      30 ITERATIONS"
6720  RETURN
6730 :
6740 :
7000  REM *********************
7010  REM * THIS SUBROUTINE    *
7020  REM * PRINTS THE RESULTS *
7030  REM *********************
7040 :
7050 :
7060 M$ = "-------------------"
7070 N$ = "------------------"
7080 M$ = M$ + N$
7090  PRINT M$
7100  FOR J = 1 TO 6
7110  FOR I = 1 TO 6
7120 F(I) = A(J,I)
7130  NEXT I
7140  GOSUB 8000
7150  PRINT G$(1); TAB( 7);G$(2);
      TAB( 14);G$(3); TAB( 21);G$
     (4); TAB( 28);G$(5); TAB( 35
     );G$(6)
7160  NEXT J
7170  PRINT M$
7180  RETURN
7190 :
7200 :
8000  REM ********************
8010  REM * THIS SUBROUTINE   *
8020  REM * SORTS SCIENTIFIC  *
8030  REM * NOTATION FOR PRINT*
8040  REM * CLARITY.          *
8050  REM ********************
8060 :
8070 :
8080  FOR I = 1 TO N
8090  IF  ABS (F(I)) > 0.01 THEN
      GOTO 8120
8100 G$(I) =  LEFT$ ( STR$ (F(I))
     ,2) +  RIGHT$ ( STR$ (F(I)),
     4)
8110  GOTO 8140
8120 G$(I) =  STR$ (F(I))
8130 G$(I) =  LEFT$ (G$(I),5)
8140  NEXT I
8150  RETURN
```

The output of this BASIC program is as follows:

```
THE ORIGINAL MATRIX IS

-------------------------------------
2.3   4.3   5.6   3.2   1.4   2.2
1.4   2.4   5.7   8.4   3.4   5.2
2.5   6.5   4.2   7.1   4.7   9.3
3.8   5.7   2.9   1.6   2.5   7.9
2.4   5.4   3.7   6.2   3.9   1.8
1.8   1.7   3.9   4.6   5.7   5.9
-------------------------------------
```

```
IN UPPER HESSENBERG FORM
-------------------------------------
2.3    10.39  11.59  5.117  2.901  2.2
3.8    10.92  14.86  2.398  7.890  7.9
00     13.61  7.327  3.631  4.041  2.289
00     00     2.443  -2.14  2.162  2.010
00     00     00     -2.47  -.890  -3.39
00     00     00     00     3.725  2.775
-------------------------------------


THE EIGENVALUES ARE
----------------------------
  REAL PART    IMAGINARY PART
----------------------------
25.5275739      0
-5.63130534     0
-.682468529     1.56595939
-.682468529     -1.56595939
.884334228      3.444546
.884334228      -3.444546
----------------------------
```

This program required less than 45 seconds to run on an Apple II computer. An Apple II computer with 48K of memory could handle a much larger sized matrix than the one considered in this example; however, it is possible that additional run time would be required. The long run time required for this problem is an indication of the relative complexity of the program and the algorithms it uses.

4.7 CONSIDERATIONS IN THE SELECTION OF AN EIGENVALUE ALGORITHM

The selection of an appropriate algorithm for a given eigenvalue problem will depend on the eigenvalue type, the matrix type, the number of eigenvalues desired, and whether the eigenvectors are required. As the complexity of the eigenvalue problem increases, the number of alternate algorithms decreases. To aid in the selection process, Table 4-1 is provided. When implementing these methods with the small computer, the overall limitations of the small computer must be kept in mind. These are as follows:

1. **Consider the nature of the problem and its solutions.** If only one eigenvalue is required, the iterative method is best to use. If all eigenvalues of a symmetric matrix are required, the Jacobi method or other transform method is best to use. If all eigenvalues of a

Table 4-1 Applications Chart for Eigenvalue Algorithm Selection

Algorithm Name (and Type)	*Applied to:*	*Result*	*Recommended When Number of Eigenvalues Desired is:* Largest or Smallest	All $\leqslant 6$	All $\geqslant 6$	*Comments*
Determinant (iteration)	General matrix	Eigenvalues		•		Requires that the roots of a general polynomial be found
Iteration (iteration)	General matrix	Eigenvalues and eigenvectors	•	•		Best accuracy for largest and smallest eigenvalues
Jacobi (transformation)	Symmetric	Diagonal form		•	•	Theoretically requires an infinite number of steps
Given's (transformation)	Symmetric	Tridiagonal		•	•	Requires roots of an easy polynomial
	Nonsymmetric	Hessenberg		•	•	Requires additional method
Householder's (transformation)	Symmetric	Tridiagonal		•	•	Requires roots of an easy polynomial
	Nonsymmetric	Hessenberg				Requires additional method
LR method (transformation)	General matrix	Block diagonal		•	•	Can be unstable
QR method (transformation)	General matrix	Block diagonal				Best general method

general matrix must be found, the *QR* algorithm must be used since the occurrence of complex eigenvalues cannot be handled by any of the other methods.

2. **Consider the computer space and run time required.** There is frequently a tradeoff between computer run time and computer storage space that will determine which method must be used for eigenvalue systems on a small computer. For example, the illustrative problems from this chapter show that the more complex methods require more computer space for the variables used in a program and a higher number of program lines. Also, if the eigenvectors are not required, the computer space required for the solution can be made smaller. In many cases the method presented in Example 4-2 could be made to run faster and with less computer storage if the parts of the program that relate to the eigenvectors were to be deleted.
3. **Consider the intermediate outputs of the computer.** Even though the process of outputting information to a CRT screen or printer may slow down an iterative solution somewhat, this intermediate information concerning the progress of the technique toward achieving a solution can often give the user considerable information as to whether the solution will ultimately converge or is failing to converge.
4. **Consider the accuracy required.** Care should be exercised in handling iterative procedures for eigenvalue extraction to ensure that maximum computational accuracy is achieved. For example, in the elimination method presented in Section 4.5, the use of row and column interchanges to find the largest pivot element is recommended.

REFERENCES

1. ARDEN, BRUCE W., and KENNETH N. ASTILL, *Numerical Algorithms: Origins and Applications*, Addison-Wesley Publishing Co., Reading, Mass., 1970.
2. BATHE, K.J., and E.L. WILSON, "Solution Methods for Eigenvalue Problems in Structural Mechanics," *International Journal of Numerical Methods in Engineering*, vol. 6, 1972, 213–226.
3. BEREZIN, I.S., and N.P. ZHIDKOV, *Computing Methods*, vol. II, Pergamon Press, Elmsford, N.Y., 1965.
4. FRANCIS, J.G.F., "The QR Algorithm, Parts I and II," *Computer Journal*, vol. 4, 1961, 265–271 and 332–345.

5. GASTINES, NOEL, *Linear Numerical Analysis*, Academic Press, Inc., New York, 1970.

6. GUPTA, K.K., "Recent Advances in Numerical Analysis of Structural Eigenvalue Problems," in *Theory and Practice in Finite Element Structural Analysis*, J.T. Oden, ed.

7. MARTIN, R.S., P. PETERS, and J.H. WILKINSON, "The QR Algorithm for Real Hessenberg Matrices," *Numerische Math.*, vol. 14, 1970, 219-231.

8. PILKEY, WALTER, and BARBARA PILKEY, eds., *Shock and Vibration Computer Programs, Reviews and Summaries*, Shock and Vibration Information Center, U.S. Department of Defense, 1975.

9. RALSTON, ANTHONY, *A First Course in Numerical Analysis*, McGraw-Hill Book Co., New York, 1965.

10. WILKINSON, J.H., *The Algebraic Eigenvalue Problem*, Oxford University Press, Inc., New York, 1965.

11. WILLIAM, P.W., *Numerical Computation*, Harper & Row, Publishers, Inc., New York, 1972.

The microcomputer coupled with a plotter can produce outstanding graphical output for scientific and engineering problem solving. (*Photo courtesy of Hewlett-Packard.*)

Ordinary differential equations

5

Any equation containing one or more derivatives is called a *differential equation*. Since most physical laws of science and engineering are expressed in terms of differential equations, the need to solve this type of equation occurs again and again. Indeed, any design analysis problem concerned with the movement of mass or energy will ultimately lead to a differential equation. Unfortunately, the number of differential equations that can be satisfactorily treated by hand calculations is quite limited. For this reason the topic of microcomputer solution of differential equations is especially important to a discussion of scientific and engineering problem solving.

Differential equations fall into two distinct categories depending on the number of independent variables they contain and thus on the type of derivatives they contain. If only one independent variable exists, the derivatives will be ordinary, and the equation will be called an *ordinary differential equation*. If more than one independent variable exists, the derivatives must be partial, and the equation will be called a *partial differential equation*. It is the purpose of this chapter to discuss solution methods for ordinary differential equations. The topic of partial differential equations will not be considered.

Throughout this chapter it will be assumed that the reader has a basic understanding of the terminology of differential equations. The reader who feels deficient in this area may wish to undertake a brief review before proceeding.

5.1 CATEGORIES OF ORDINARY DIFFERENTIAL EQUATIONS

The solution of problems involving differential equations requires that the dependent variable and/or its derivatives be given at prescribed values of the independent variable. Whenever all these constraints are imposed at the zero value of the independent variable, the solution task is said to be an *initial-value problem*. Whenever any of the constraints are imposed at any independent variable value other than zero, the solution task is said to be a *boundary-value problem*. Constraints on initial-value problems are generally termed *initial conditions*, and the constraints on boundary value problems are termed *boundary conditions*. The most common type of independent variable used in an initial-value problem is time. For example, the free vibration of a spring-supported mass would have its displacement described by a differential equation using time as the independent variable. If the constraints on the motion are given for the displacement and velocity at time equal to zero, the problem would be categorized as an initial-value problem. This same device would give rise to a boundary-value problem if one of the constraints involved a given displacement at the end of a prescribed time interval. Frequently, a boundary-value problem has length as its independent variable rather than time. A familiar example of this situation is the differential equation that describes the deflection of a beam. In this situation, boundary conditions are usually specified at the two ends. Although these two categories of problems are being discussed in the same chapter, the computational differences in their numerical solutions are substantial.

Initial-Value Problems

The initial-value problem can be stated in simple terms. Given an initial condition $y(x_0) = y_0$ and the differential equation

$$\frac{dy}{dx} = f(x, y)$$

find the unknown function $y(x)$ that satisfies both the differential equation and the initial condition. Generally, a numerical solution to this problem is accomplished by first computing the slope of the curve and then proceeding in small increments of x to a new point $x_1 = x_0 + h$. The new point is obtained using information about the slope of the curve as calculated from the differential equation. Thus, the numerical solution is composed of a series of short, straight-line segments that approximate the true curve of $y = f(x)$. The numerical method itself is concerned with the calculation procedure for moving from point to point on the curve.

Since the topic of numerical solution of initial-value problems is of such importance to many fields of science and engineering, this topic has received considerable attention over the years. As a result, an extremely large number of methods exist. No attempt will be made to discuss every such algorithm; rather, it is intended to discuss the most common of those that represent the following two categories of initial-value problems.

1. *One-step methods* are methods that use information about a single, previous step to find the next point on a curve $y = f(x)$. Methods of this type include Euler's method and the Runge–Kutta methods.
2. *Predictor–corrector* methods, also known as multistep methods, find the next point on a curve $y = f(x)$ by using information from more than one previous point. Iteration is often used to achieve a sufficiently close numerical value. Methods of this type include Milne's method, the Adams–Bashforth method, and the Hamming method.

Errors

Before discussing the specific differential equation methods, it is important to understand the source of errors associated with the numerical approximations. Three sources of errors occur.

1. *Round-off error* is due to the numerical limitations of the computer being used. Every computer has a limitation on the number of significant digits that it can store and manipulate.
2. *Truncation error* is due to the fact that infinite series used to approximate a function are often truncated after a few terms. Such a procedure is commonplace in numerical methods and introduces errors that are due entirely to the method rather than the machine.
3. *Propagation error* is due to the accumulation of previous errors in a numerical scheme. Any approximate technique is never exact. Thus, any error that an approximation scheme introduces at an early step will be carried along in the computation process for later steps.

These three sources of errors give rise to two types of observed errors.

1. *Local error* is the amount of error that enters the computational process at any given computational step.
2. *Global error* at any point in the computation is the difference between the computed value of the solution and the exact solution. Thus the global error accounts for the total accumulation of error from the start of the computational process.

5.2 ONE-STEP METHODS FOR THE INITIAL-VALUE PROBLEM

The one-step methods may be used to solve a first-order differential equation of the form

$$y' = f(x, y)$$

where y' represents dy/dx. The initial condition for this problem is expressed as $y(x_0) = y_0$. The purpose of a one-step method is to provide a means for calculating a sequence of y values corresponding to discrete values of the independent variable.

Euler's Method

Euler's method is the simplest of all methods for solving the initial-value problem. It is applicable to first-order differential equations and has rather limited accuracy. Because of this characteristic, it is not recommended for practical use. It does, however, provide considerable insight into the understanding of several other, more practical methods.

The basis for the Euler method comes from an application of the initial conditions to a Taylor's series of the form

$$y(x_0 + h) = y(x_0) + hy'(x_0) + (\tfrac{1}{2})h^2 y''(x_0) + \cdots$$

If the value of h is quite small, terms containing powers of h^2 and higher will be much smaller and will be neglected. When this is done, the expansion becomes

$$y(x_0 + h) = y(x_0) + hy'(x_0)$$

Evaluation of the differential equation at the initial condition will give $y'(x_0)$. Thus the dependent variable can be approximated at a small step h away from the initial point. This process can be continued for as many steps as desired using the relationship

$$y_{n+1} = y_n + hf(x_n, y_n), \qquad n = 1, 2, \ldots$$

The Euler method is presented in graphical form in Figure 5-1. The error in this method is said to be on the order of h^2 since terms containing h^2 and higher were neglected in the original Taylor's expansion.

Modified Euler Method

One difficulty associated with the use of Euler's method is that, although the slope of the exact curve at the starting point is $y'(x_0)$, this slope changes as the independent variable changes. Thus at the point $x_0 + h$ the slope is no longer the same as it was at the start of the interval. Thus, an error is introduced into the calculation process whenever the starting slope is used for the whole interval. The accuracy of the Euler method can be substantially improved if a better approximation is used for the derivative. One possible way to do this is to use an average value of the derivatives at the beginning and the end of the interval. The modified Euler method does this by taking a temporary Euler step

$$y^*_{n+1} = y_n + hf(x_n, y_n)$$

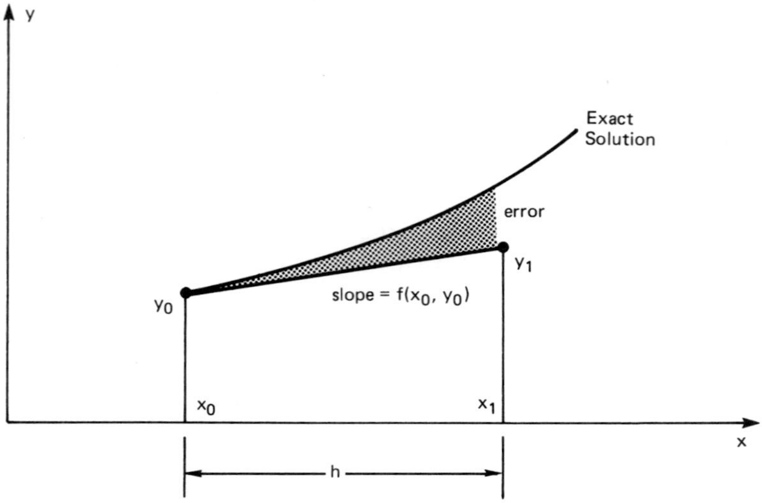

Figure 5-1 The Euler method.

and then using y^*_{n+1} to calculate an approximation to the derivative at the end of the interval. This derivative will be $f(x_{n+1}, y^*_{n+1})$. This new derivative is averaged with the initial derivative to obtain a more accurate value for y_{n+1}:

$$y_{n+1} = y_n + \tfrac{1}{2}h(f(x_n, y_n) + f(x_{n+1}, y^*_{n+1}))$$

This procedure is illustrated in Figure 5-2. Another way to understand how the modified Euler method is developed is to return to the Taylor's

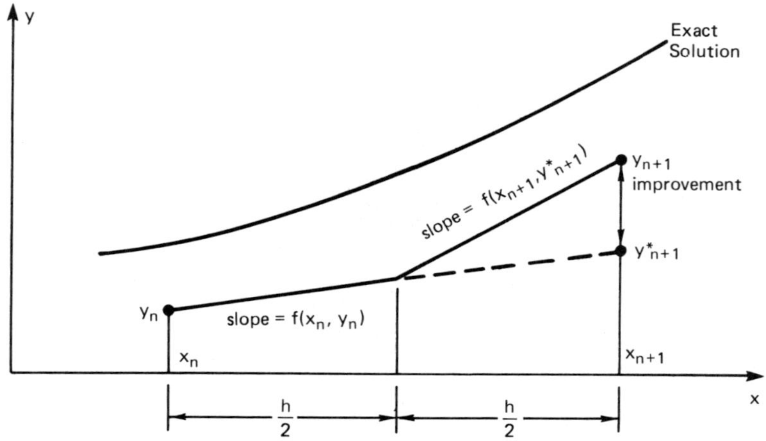

Figure 5-2 The modified Euler method.

expansion

$$y(x_0 + h) = y(x_0) = hy'(x_0) + \frac{h^2}{2} y''(x_0) + \cdots$$

If the h^2 terms are retained and all higher-order terms are dropped, it seems reasonable to expect more accuracy. It should be noted, however, that saving the h^2 term requires that the second derivative $y''(x_0)$ be known. This second derivative can be approximated by a finite difference:

$$y''(x_0) = \frac{\Delta y'}{\Delta x} = \frac{y'(x_0 + h) - y'(x_0)}{h}$$

If this expression is substituted into the truncated Taylor's expansion, the result will be

$$y(x_0 + h) = y(x_0) + (\tfrac{1}{2})h(y'(x_0 + h) + y'(x_0))$$

which is equivalent to the expression found previously.

This method is called a *second-order method* since it utilizes the h^2 terms in the Taylor's expansion. The error at each step is on the order of h^3. The price one pays to achieve this improved accuracy is the time spent in the extra calculation to get y^*_{n+1}. This method suggests that additional accuracy could be achieved if the user were willing to invest the time to achieve an even better approximation for the derivative (i.e., if the user were willing to include additional terms in the Taylor's expansion). This concept is the basis for the Runge–Kutta methods.

Runge–Kutta Methods

To retain the nth-order term in a Taylor's expansion, it is necessary to be able to evaluate the nth-order derivative of the dependent variable. In the modified Euler method, the slopes at the two ends of the interval of interest were sufficient to provide a finite difference form for the second derivative. To use a finite difference form to calculate the third derivative, it will be necessary to know the second derivative at at least two different locations. This requires that an additional slope be evaluated at an intermediate point within the h interval from x_n to x_{n+1}. Indeed, the higher the order of the derivative that must be calculated, the higher will be the number of additional, internal evaluations that must be made. The Runge–Kutta method provides a set of formulas for selecting the spacing of the internal evaluations required to implement this strategy. Since a number of alternatives exist for the spacing and for the relative weighting to be used for the slopes found, the term Runge–Kutta refers to a large family of methods for handling first-order differential equations. The most commonly used Runge–Kutta formulation is based on retaining all terms up through h^4, and is thus termed a *fourth-order* method having an error at any step on the order of h^5. The calculation formula for this classical method is

$$y_{n+1} = y_n + \frac{(K_0 + 2K_1 + 2K_2 + K_3)}{6}$$

where

$$K_0 = hf(x_n, y_n)$$
$$K_1 = hf(x_n + 0.5h, y_n + 0.5K_0)$$
$$K_2 = hf(x_n + 0.5h, y_n + 0.5K_1)$$
$$K_3 = hf(x_n + h, y_n + K_2)$$

In reality, the Euler and modified Euler methods are actually first- and second-order Runge–Kutta methods. The increased accuracy made possible by the Runge–Kutta technique makes it far more desirable to use than either of the previous two methods discussed, and more than justifies the additional computational effort needed to use them. Because of this greater accuracy, it is often possible to use a larger-sized h interval. The allowable error at each step will determine the maximum allowable step size that can be used. It is wise to adjust the value of h to the maximum allowable in order to achieve optimum efficiency in the computational process. Frequently, this adjustment process is included as an automatic part of a computational scheme employing the Runge–Kutta method.

The best way to illustrate the relative accuracy of the one-step methods is by means of an example.

EXAMPLE 5-1 Suppose it is desired to find the solution to the equation

$$\frac{dy}{dx} = 2x^2 + 2y$$

subject to $y(0) = 1$ on the interval $0 \leqslant x \leqslant 1$ with $h = 0.1$. Since this problem is linear, the exact solution is known to be

$$y = 1.5e^{2x} - x^2 - x - 0.5$$

and can be used to inspect the relative accuracy of the various methods. A comparison of the results is presented in the following table. Clearly, the Runge–Kutta method is better than the Euler method or the modified Euler method.

X_n	*Euler*	*Modified Euler*	*Runge-Kutta*	*Exact*
0.0	1.0000	1.0000	1.0000	1.0000
0.1	1.2000	1.2210	1.2221	1.2221
0.2	1.4420	1.4923	1.4977	1.4977
0.3	1.7384	1.8284	1.8432	1.8432
0.4	2.1041	2.2466	2.2783	2.2783
0.5	2.5569	2.7680	2.8274	2.8274
0.6	3.1183	3.4176	3.5201	3.5202
0.7	3.8139	4.2257	4.3927	4.3928
0.8	4.6747	5.2288	5.4894	5.4895
0.9	5.7376	6.4704	6.8643	6.8645
1.0	7.0472	8.0032	8.5834	8.5836

Runge–Kutta Methods for a System of Differential Equations

Any of the Runge–Kutta formulas can be used to solve simultaneous differential equations, and thus higher-order differential equations, since one can make n first-order differential equations from a single nth-order differential equation. For example, in the second-order differential equation

$$\frac{d^2y}{dx^2} = g\left(x, y, \frac{dy}{dx}\right)$$

one can let

$$z = \frac{dy}{dx}$$

and then

$$\frac{dz}{dx} = \frac{d^2y}{dx^2}$$

The two first-order equations become

$$\frac{dz}{dx} = g(x, y, z) \quad \text{and} \quad \frac{dy}{dx} = f(x, y, z)$$

where in this case $f(x, y, z) = z$.

The initial-value problem for this situation would be specified in terms of two initial conditions:

$$y(x_0) = y_0 \quad \text{and} \quad z(x_0) = z_0$$

The Runge–Kutta formulas for this problem would be

$$y_{n+1} = y_n + K$$

and

$$z_{n+1} = z_n + L$$

where

$$K = \frac{K_1 + 2K_2 + 2K_3 + K_4}{6}$$

$$L = \frac{L_1 + 2L_2 + 2L_3 + L_4}{6}$$

In these equations,

$$K_1 = hf(x_n, y_n, z_n)$$
$$L_1 = hg(x_n, y_n, z_n)$$
$$K_2 = hf(x_n + 0.5h, y_n + 0.5K_1, z_n + 0.5L_1)$$
$$L_2 = hg(x_n + 0.5h, y_n + 0.5K_1, z_n + 0.5L_1)$$
$$K_3 = hf(x_n + 0.5h, y_n + 0.5k_2, z_n + 0.5L_2)$$
$$L_3 = hg(x_n = 0.5h, y_n + 0.5K_2, z_n + 0.5L_2)$$
$$K_4 = hf(x_n + h, y_n + K_3, z_n + L_3)$$
$$L_4 = hg(x_n + h, y_n + K_3, z_n + L_3)$$

Gill's Modification of the Runge–Kutta Method

The Runge–Kutta method as formulated in the previous section requires further modifications as the number of simultaneous equations increases. It also does not handle the round-off errors in a way so as to minimize their effect on the overall result. To provide a method that overcomes these disadvantages, Gill developed a calculation procedure that has the following special features:

1. It requires a minimum number of storage spaces,
2. It gives the highest attainable accuracy in terms of round-off errors.
3. It requires only a small number of computer instructions to implement.

For these reasons it is well suited for use on the microcomputer. Gill's procedure applies to a system of $n + 1$ first-order equations of the form

$$y_i'(x) = f_i(x, y_0(x), y_1(x), \ldots, y_n(x)), \qquad i = 0, 1, 2, \ldots, n$$

The method handles the values of y in a two-dimensional array $y_{i,j}$, where the initial conditions are expressed as

$$y_{io} = y_i(x_0), \qquad \text{for } i = 0, 1, 2, \ldots, n$$

To start the method, a special set of coefficients must be loaded as follows:

$$\begin{array}{lll} a_1 = \frac{1}{2} & b_1 = 2 & c_1 = \frac{1}{2} \\ a_2 = 1 - \sqrt{\frac{1}{2}} & b_2 = 1 & c_2 = 1 - \sqrt{\frac{1}{2}} \\ a_3 = 1 + \sqrt{\frac{1}{2}} & b_3 = 1 & c_3 = 1 + \sqrt{\frac{1}{2}} \\ a_4 = \frac{1}{6} & b_4 = 2 & c_4 = \frac{1}{2} \end{array}$$

A register q_{ij} will be used with

$$q_{io}(x_0) = 0, \qquad \text{for } i = 0, 1, \ldots, n$$

The procedure starts by setting the index $j = 1$. Next the value of

$$k_{ij} = f_i(x_{j-1}, y_{0,j-1}, y_{1,j-1}, \ldots, y_{n,j-1})$$

is calculated for $i = 0, 1, \ldots, n$. Next the values of

$$y_{ij} = y_{i,j-1} + [a_j(hk_{ij} - b_j q_{i,j-1})]$$

and

$$q_{ij} = q_{i,j-1} + 3[a_j(hk_{ij} - b_j q_{i,j-1})] - c_j hk_{ij}$$

are calculated for $i = 0, 1, \ldots, n$. The procedure from the step k_{ij} above is repeated for $j = 2, 3,$ and 4. As the process proceeds, the values of x_i are changed as follows:

$$x_1 = x_0 + \frac{h}{2}$$

$$x_2 = x_0 + \frac{h}{2}$$

$$x_3 = x_0 + h$$

At the end of this four-stage process, the values of y_i at $x = x + h$ will be

$$y_i(x + h) = y_{i4}, \qquad i = 0, 1, \ldots, n$$

The initial conditions y_{io} and q_{io} are updated as follows:

$$\left.\begin{array}{l} y_{io} = y_{i4} \\ q_{io} = q_{i4} \end{array}\right. , \qquad \text{for } i = 0, 1, \ldots, n$$

The whole process starting with the step $j = 1$ above is repeated to find subsequent values of y_i.

To allow control over accuracy as the method progresses, the usual procedure is to perform two Gill steps of size h to get $y_i(x + 2h) = y_i^{(1)}$, and then to perform one Gill step of size $2h$ to get $y_i(x + 2h) = y_i^{(2)}$. Since the computation involving the two smaller steps should give greater accuracy, a comparison of these two results at $x + 2h$ should provide a measure of the local truncation error. If the difference in the two values is smaller than some prescribed value ϵ, the result is said to be sufficiently accurate. If the difference is larger than the prescribed value, the accuracy can be improved by decreasing the step size by one-half and repeating the process. For this process, the truncation error is

$$\text{Error} = \frac{1}{15} \sum_{i=0}^{n} \frac{|y_i^{(1)} - y_i^{(2)}|}{n + 1}$$

Thus if Error $> \epsilon$, then $h = h/2$ and the calculations are repeated to find an answer with more acceptable accuracy prior to going on to the next step. This step-size reduction can be repeated as many times as is necessary to achieve the desired degree of accuracy. Frequently, Gill's method is implemented with provisions for a doubling of the step size if the error is small enough to justify this procedure. For example, if Error $< \epsilon/50$, then h is replaced with $2h$. This procedure will speed the solution process.

An example of the use of Gill's modification of the Runge–Kutta method will be presented, but first let us summarize the general characteristics of all one-step methods.

A Summary of One-Step Methods

Certain special characteristics are common to all one-step methods.

1. They need only information at the preceding point (x_n, y_n, z_n) to find information at the new point $(x_{n+1}, y_{n+1}, z_{n+1})$. This characteristic is known as *self-starting behavior*.
2. They all stem from a Taylor's expansion of the function and contain terms up to and including the h^k term. The integer k is called the *order* of the method, and the error at any step is said to be on the order of $k + 1$.
3. They do not require the actual evaluation of any derivatives but only of the function itself. They may, however, require several evaluations of the function at intermediate points. This, of course, can be time and effort consuming.
4. Their self-starting nature lends itself to easy changes in the step size h.

To illustrate the one-step methods, let us now consider an example application involving Gill's modification of the Runge–Kutta method.

EXAMPLE 5-2

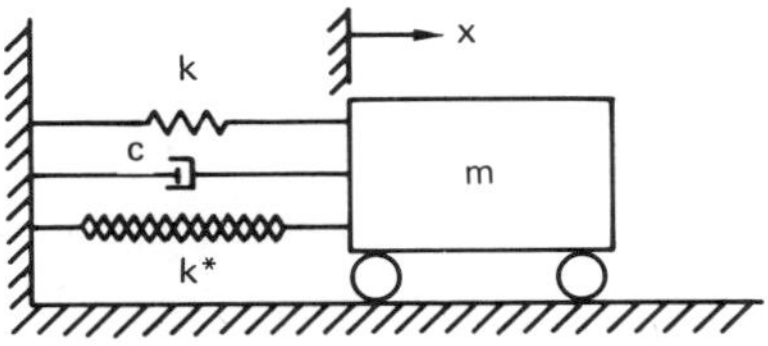

Shock and vibration problems in the aerospace and transportation industries arise from many different types of excitation sources. The elimination of shock and vibration is of crucial importance in the isolation of instruments and controls or in the protection of human occupants of vehicles. The usual solution to problems involving excess vibration transmission involves the use of lightly damped flexible supports. These soft supports cause the natural frequency of a suspension system to be far below the disturbing frequency. This solution is effective for the isolation of steady-state vibration; however, when these suspensions encounter shock excitation, their softness often leads to damagingly large deflections. It has been pointed out that this undesirable feature is not present in suspension systems utilizing symmetrically nonlinear springs that harden.[1] These springs become progressively stiffer when subjected to large deflections from the operating point. The device shown consists of a mass m connected to the rigid wall by means of a linear spring with coefficient k, a damper with coefficient c, and a nonlinear spring that exerts a restoring force proportional to a constant k^* times the third power of displacement. This "cubic" spring will provide a symmetrically nonlinear behavior to satisfy the need for shock and vibration isolation.

Since the resulting differential equation for the motion of this system is described by the nonlinear differential equation

$$m\ddot{x} + c\dot{x} + kx + k^*x^3 = 0$$

the displacement x as a function of time cannot be found by traditional "exact" methods. For this reason a numerical solution to this differential equation is desirable.

If the physical parameters of the suspension system are

$$k = 2.0 \text{ N/cm}$$
$$k^* = 2.0 \text{ N/cm}^3$$
$$c = 0.15 \text{ N sec/cm}$$
$$m = 1.0 \text{ kg}$$

and the initial conditions are

$$x(0) = 10. \text{ cm}$$
$$\dot{x}(0) = 0. \text{ cm/sec}$$

prepare and run a computer program that will simulate the motion of this system for time from 0 to 1.0 seconds.

[1] J. A. Tobias, "Design of Small Isolator Units for the Suppression of Low-Frequency Vibration," *Journal of Mechanical Engineering Science*, vol. 1, no. 3, 1959, pp. 280–292.

To solve this problem by a one-step method, it will be necessary to reduce the second-order differential equation to two first-order differential equations. When this is done, the result will be

$$\dot{x}_1 = x_2$$

$$\dot{x}_2 = \frac{-c}{m}\dot{x}_1 - \frac{k}{m}x_1 - \frac{k^*}{m}x_1^3$$

To utilize centimeter dimensions in the output, it is necessary to convert the mass to

$$m = 0.01 \text{ N sec}^2/\text{cm}$$

A BASIC program that solves this problem now follows. This program utilizes the Runge–Kutta method with Gill's modification. The program also utilizes the step-size adjustment procedure described previously in order to move through to the final value with a minimum of time for a given accuracy. The differential equation is loaded by a special subroutine starting at statement 5000, and the output is printed by a special subroutine starting at statement 6000. The actual use of Gill's procedure is accomplished by the subroutine starting at statement 4800. Since the overall procedure calls for a comparison of results gathered by changing step sizes, several working arrays are required to store the intermediate results.

```
1000  REM **********************
1010  REM *THIS PROGRAM FINDS   *
1020  REM *THE SOLUTION OF A    *
1030  REM *DIFFERENTIAL EQUA-   *
1040  REM *TION BY THE RUNGE-   *
1050  REM *KUTTA METHOD.        *
1060  REM **********************
1070 :
1080 :
1090 :
1100  DIM Y(2),DERY(2)
1110 X = 0:XLAST = 1
1120 Y(0) = 10.:Y(1) = 0.
1130 H = 0.02:N = 1
1140 EPS = .001
1150 :
1160  PRINT "-------------------
 -----------------"
1170  PRINT "     X            Y
            Y'"
1180  PRINT "-------------------
 -----------------"
1190  GOSUB 3000
1200  PRINT "-------------------
 -----------------"
1210  END
1220 :
1230 :
3000  REM **********************
3010  REM * THIS SUBROUTINE     *
3020  REM * SOLVES A SYSTEM OF  *
3030  REM * FIRST ORDER DIFFER-*
3040  REM * ENTIAL EQUATIONS    *
3050  REM * WITH GIVEN INITIAL  *
3060  REM * VALUES.             *
3070  REM *                     *
3080  REM * THE METHOD USED IS  *
3090  REM * A FOURTH ORDER      *
3100  REM * RUNGE-KUTTA FORMULA*
3110  REM * WITH GILL'S MODIFI-*
3120  REM * CATION.             *
3130  REM *                     *
3140  REM * STEPSIZE IS AUTO-   *
3150  REM * MATICALLY ADJUSTED *
```

```
3160  REM * BY DOUBLING OR       *
3170  REM * HALVING TO MAINTAIN*
3180  REM * THE DESIRED DEGREE *
3190  REM * OF ACCURACY.         *
3200  REM *                      *
3210  REM *  N+1      - THE NUM- *
3220  REM *             BER OF   *
3230  REM *             EQUATIONS*
3240  REM *             IN THE   *
3250  REM *             SYSTEM.  *
3260  REM *                      *
3270  REM *  X        - INDEPEN- *
3280  REM *             DENT VAR-*
3290  REM *             IABLE.   *
3300  REM *                      *
3310  REM * Y(I)      - VECTOR OF*
3320  REM *             FUNCTION *
3330  REM *             VALUES.  *
3340  REM *                      *
3350  REM * XLAST     - FINAL    *
3360  REM *             VALUE OF *
3370  REM *             INDEPEN- *
3380  REM *             DENT     *
3390  REM *             VARIABLE.*
3400  REM *                      *
3410  REM *                      *
3420  REM * DERY(I) - VECTOR OF*
3430  REM *             DERIVA-  *
3440  REM *             TIVES.   *
3450  REM *                      *
3460  REM *  H        - INITIAL  *
3470  REM *             STEP SIZE*
3480  REM *                      *
3490  REM * EPS       - ERROR    *
3500  REM *             VALUE FOR*
3510  REM *             TESTING  *
3520  REM *             TRUN-    *
3530  REM *             CATION.  *
3540  REM *                      *
3550  REM * 5000      - SUBROU-  *
3560  REM *             TINE TO  *
3570  REM *             LOAD THE *
3580  REM *             DERIVA-  *
3590  REM *             TIVES    *
3600  REM *             FROM THE *
3610  REM *             EQUATIONS*
3620  REM *                      *
3630  REM * 6000      - SUBROU-  *
3640  REM *             TINE TO  *
3650  REM *             PRINT    *
3660  REM *             OUTPUT.  *
3670  REM *                      *
3680  REM **********************
3690 :
3700  REM *INITIALIZE PARAMETERS
3710  REM *AND DIMENSION
3720  REM *WORKING ARRAYS.
3730  DIM A(5),B(5),C(5)
3740  DIM Q(N),YY(3,N),DD(3,N),QQ
    (3,N)
3750 A(1) = .5:A(4) = 1 / 6
3760 A(3) = 1 +  SQR (.5)
3770 A(2) = 1 -  SQR (.5)
3780 B(1) = 2:B(2) = 1.
3790 B(3) = 1:B(4) = 2.
3800 C(1) = .5:C(2) = A(2)
3810 C(3) = A(3):C(4) = .5
3820 HMAX = H
3830  FOR I = 0 TO N
3840 Q(I) = 0.
3850 YY(0,I) = Y(I)
3860 QQ(0,I) = Q(I)
3870  NEXT I
3880  GOSUB 5000: GOSUB 6000
3890 :
3900  REM **K=0, DO 1ST H STEP
3910 K = 0
3920  GOSUB 4800
3930 :
3940  REM **1ST H STEP COMPLETE
3950  REM **STORE RESULTS
3960 K = 1
3970  FOR I = 0 TO N
3980 YY(K,I) = Y(I)
3990 DD(K,I) = DERY(I)
4000 QQ(K,I) = Q(I)
4010  NEXT I
4020 :
4030  REM **DO 2ND H STEP
4040  GOSUB 4800
4050 :
4060  REM **2ND H STEP COMPLETE
4070  REM **STORE RESULTS
4080 K = 2
4090  FOR I = 0 TO N
4100 YY(K,I) = Y(I)
4110 DD(K,I) = DERY(I)
4120 QQ(K,I) = Q(I)
4130 Y(I) = YY(0,I)
4140 Q(I) = QQ(0,I)
4150  NEXT I
4160 :
4170  REM **RESET PARAMETERS
4180  REM **AND DO DOUBLE H STEP
4190 X = X - 2 * H
4200 H = H + H
4210  GOSUB 4800
4220 :
4230  REM **DOUBLE H STEP DONE
4240  REM **CALCULATE ERROR SUM
4250 X = X - H:SUM = 0:H = H / 2
4260  FOR I = 0 TO N
4270 SUM = SUM +  ABS (Y(I) - YY(
    2,I)) / (N + 1)
4280  NEXT I
4290 :
4300  REM **IF ERROR TOO LARGE
4310  REM **HALVE H AND RETRY
4320  IF SUM / 15 > EPS THEN  GOTO
    4580
```

```
4330 :
4340  REM **IF ERROR IS SMALL
4350  REM **ENOUGH, PRINT
4360  REM **RESULTS AT H & 2H
4370  FOR J = 1 TO 2
4380  FOR I = 0 TO N
4390 Y(I) = YY(J,I)
4400 DERY(I) = DD(J,I)
4410  NEXT I
4420 X = X + H
4430  GOSUB 6000
4440  NEXT J
4450 :
4460  REM **STOP AFTER LAST STEP
4470  IF X > = XLAST THEN  RETURN

4480 :
4490  REM **ERROR TOO SMALL SO
4500  REM **DOUBLE STEP SIZE
4510  REM **UP TO HMAX.
4520  IF SUM / 15 > EPS / 50 THEN
      GOTO 4650
4530 H = H * 2
4540  IF H > HMAX THEN H = HMAX
4550  GOTO 4650
4560 :
4570  REM **CUT H BY HALF & REDO
4580 H = H / 2.
4590  FOR I = 0 TO N
4600 Y(I) = YY(0,I)
4610 Q(I) = QQ(0,I)
4620  NEXT I
4630  GOTO 3910
4640 :
4650  REM **PROCEED WITH
4660  REM **NEXT STEPS
4670  FOR I = 0 TO N
4680 YY(0,I) = YY(2,I)
4690 QQ(0,I) = QQ(2,I)
4700 Y(I) = YY(0,I)
4710 Q(I) = QQ(0,I)
4720  NEXT I
4730 :
4740  REM **INSURE THAT LAST
4750  REM **TWO STEPS REACH
4760  REM **THE XLAST VALUE.
4770  IF X + 2 * H > XLAST THEN H
      = (XLAST - X) / 2
4780  GOTO 3910
4790 :
4800  REM **********************
4810  REM * THIS SUBROUTINE    *
4820  REM * PERFORMS THE GILL  *
4830  REM * PROCEDURE.         *
4840  REM **********************
4850 :
4860  FOR J = 1 TO 4
4870  FOR I = 0 TO N
4880  GOSUB 5000
4890  NEXT I
4900  FOR I = 0 TO N
4910 Y(I) = Y(I) + (A(J) * (H * D
     ERY(I) - B(J) * Q(I)))
4920 Q(I) = Q(I) + 3 * (A(J) * (H
      * DERY(I) - B(J) * Q(I))) -
     C(J) * H * DERY(I)
4930  NEXT I
4940  IF J = 1 OR J = 3 THEN X =
     X + .5 * H
4950  NEXT J
4960  RETURN
4970 :
4980 :
5000  REM **********************
5010  REM *THIS SUBROUTINE     *
5020  REM *LOADS THE SYSTEM OF *
5030  REM *DIFFERENTIAL EQUA-  *
5040  REM *TIONS.              *
5050  REM **********************
5060 :
5070 :
5080 YM = .01:CC = .15
5090 YK = 2:YS = .2
5100 :
5110 DERY(0) = Y(1)
5120 DERY(1) =  - CC * Y(1) / YM -
     YK * Y(0) / YM - YS * Y(0)  ^
     3 / YM
5130  RETURN
5140 :
6000  REM **********************
6010  REM *THIS SUBROUTINE     *
6020  REM *PRINTS THE RESULTS  *
6030  REM *X AND Y(I) FOR THOSE*
6040  REM *VALUES DESIRED.     *
6050  REM **********************
6060 :
6070 :
6080  PRINT X; TAB( 13);Y(0); TAB(
      26);Y(1)
6090 :
6100  RETURN
```

An error test value of $\epsilon = 0.001$ is selected for this example. An observation of the printout provided by this computer program illustrates how the time increment is adjusted in order to maintain the desired error in an efficient fashion. The output of this computer program now follows:

X	Y	Y'
0	10	0
2.5E-03	9.93232289	-53.6344463
5E-03	9.73515263	-103.300834
7.50000001E-03	9.42044762	-147.442476
.01	9.00345207	-185.003058
.0125	8.50136574	-215.475124
.015	7.93200059	-238.86479
.0175	7.31260692	-255.596071
.02	6.65898269	-266.385094
.0225	5.98490535	-272.111861
.025	5.30186605	-273.708997
.0275000001	4.61904961	-272.07728
.0300000001	3.94348968	-268.029784
.0325000001	3.28033098	-262.261267
.0350000001	2.63314234	-255.337094
.0400000001	1.39498241	-239.658355
.0450000001	.23778144	-223.190235
.0500000001	-.837028911	-206.750077
.0550000001	-1.8296515	-190.246974
.0600000002	-2.73839987	-173.074755
.0650000001	-3.5579422	-154.429744
.0700000002	-4.27900206	-133.579432
.0750000002	-4.88930931	-110.093498
.0800000002	-5.37558268	-84.0203857
.0850000002	-5.72617951	-55.9705894
.0900000002	-5.9338456	-27.0691075
.0950000002	-5.997878	1.22866723
.1	-5.92510854	27.4131487
.105	-5.72944813	50.2048123
.11	-5.4301759	68.7599964
.115	-5.04950952	82.7488213
.12	-4.61010778	92.3090077
.125	-4.13302757	97.9190898
.13	-3.63640045	100.245972
.135	-3.13484171	100.009709
.14	-2.63944347	97.8870411
.145	-2.15814424	94.4566284
.15	-1.69628325	90.1782199
.155	-1.25719749	85.3945328
.165	-.45390522	75.1883875
.175	.24632528	64.8780732
.185	.844332359	54.7567572
.195	1.34204573	44.8073317
.205	1.74084667	34.9695919
.215	2.04190782	25.2756755
.225	2.24738225	15.8896829
.235	2.3616279	7.07970683
.245	2.39193094	-.849413774
.255	2.34848903	-7.63186863
.265	2.24371089	-13.0981789
.275	2.09110381	-17.1991835

```
.285        1.90409143   -19.997137
.295        1.69503982   -21.6353384
.305        1.47463306   -22.3008173
.315        1.2516076    -22.1913378
.335        .823187163   -20.3641961
.355        .445057678   -17.3172906
.375        .133591571   -13.7912377
.395        -.106376985  -10.2226149
.415        -.276827104  -6.87364631
.435        -.383934984  -3.9103862
.455        -.436534388  -1.43544506
.475        -.444968326  .501969498
.495        -.420071818  1.90084735
.515        -.37228445   2.80001221
.535        -.310969527  3.26612479
.555        -.243987182  3.38079817
.575        -.177510644  3.22950623
.595        -.116034551  2.89356501
.615        -.062512277  2.44517767
.635        -.0185670062 1.94488805
.655        .0152639942  1.44069447
.675000001  .0392789628  .968243621
.695000001  .0543726339  .551727857
.715000001  .0618201408  .205246307
.735000001  .0630912892  -.0655457627
.755000001  .059699422   -.261651485
.775000001  .0530861866  -.388949109
.795000001  .0445409488  -.456527779
.815000001  .0351515542  -.475228426
.835000001  .0257817671  -.456440999
.855000001  .0170700214  -.41117688
.875        9.44401193E-03-.3494096
.895000001  3.14598632E-03-.279659083
.915000001  -1.7357945E-03-.208784067
.935000001  -5.23319024E-03-.141942377
.955000001  -7.46485593E-03-.0826780348
.975000001  -8.60559987E-03-.0330963403
.995000001  -8.85982919E-035.90779659E-03
.997500001  -8.83986793E-03.0100337657
1           -8.80979719E-03.0139955672
------------------------------------
```

This output required less than 5 minutes to complete on an Apple II computer. Since the program contains extra evaluations in order to test the error performance, it naturally takes longer to run than a similar program without this capability. The reader will also note that the program contains special provisions to guarantee that the final calculation is performed on the last desired value of the independent variable.

A plot of the displacement portion of these data follows. From this figure it is obvious that the frequency of vibration is dependent on the amplitude of the vibration. This behavior is characteristic of nonlinear systems that contain springs that harden as they deflect.

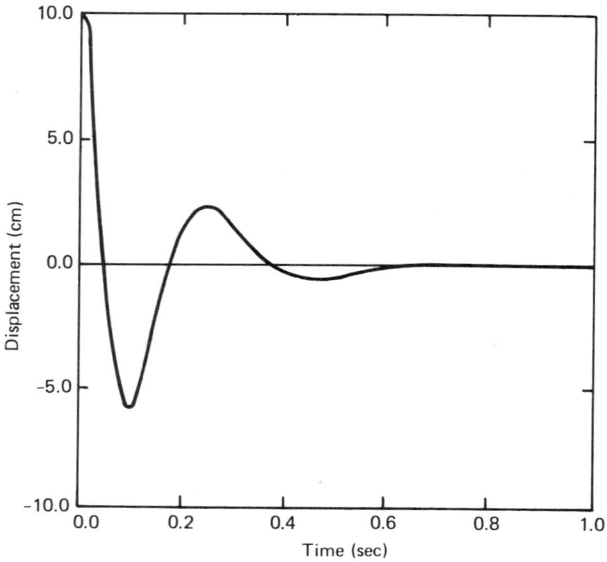

5.3 PREDICTOR-CORRECTOR METHODS

Predictor–corrector methods use information from several previous points to compute a new point. They make use of two formulas, one called a predictor equation and the other called a corrector equation. Regardless of the predictor–corrector method used, the logic flow will be the same. Only the prediction and correction formulas are changed to change methods. The logic of this differential equation technique is presented in Figure 5-3 for the solution of the equation

$$y'(x) = f(x, y)$$

Since predictor–corrector methods require information about several previous points in order to move forward, they are not self-starting, as were the one-step methods. For this reason, all predictor–corrector methods require the use of a one-step method to get started. A Runge–Kutta method is frequently employed to meet this need. The method then proceeds as follows. First, the prediction formula is applied to

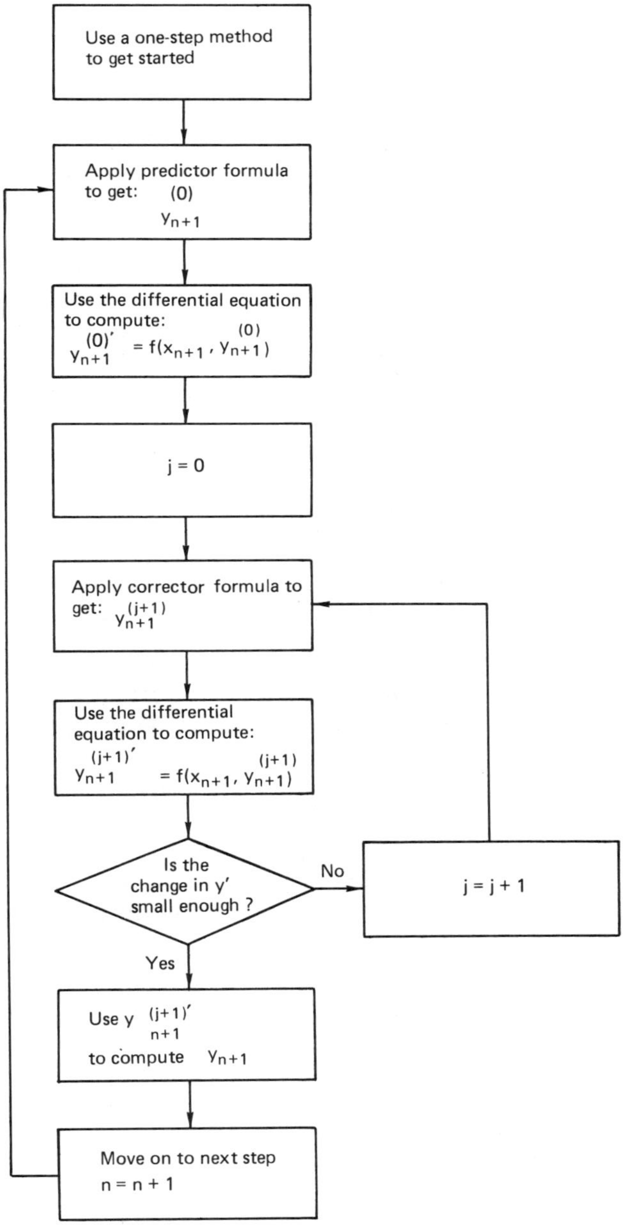

Figure 5-3 The predictor-corrector method.

the starting information to predict a value for

$$y_{n+1}^{(0)}$$

The superscript (0) indicates that this prediction is just one of a sequence of y_{n+1} values that are better and better in terms of accuracy. Using this initial projection for y_{n+1}, the differential equation is employed to compute the derivative.

$$y_{n+1}^{(0)\prime} = f(x_{n+1}, y_{n+1}^{(0)})$$

Once this derivative is available, it is used in the corrector formula to obtain an improved value:

$$y_{n+1}^{(j+1)}$$

This value is, in turn, used to improve the derivative by means of the differential equation

$$y_{n+1}^{(j+1)\prime} = f(x_{n+1}, y_{n+1}^{(j+1)})$$

If this derivative is not sufficiently close to the previous one, the new value is applied to the corrector formula to continue the iterative improvement process. If, on the other hand, the change in derivatives is small enough, the value of

$$y_{n+1}^{(j+1)\prime}$$

is used to compute and document the final value of y_{n+1}. Once this has been achieved, the process may be repeated to move onward to the next step point y_{n+2}.

The usual approach to the derivation of predictor or corrector formulas is to view the stepping problem as if it were the process of approximate integration and thus to use finite-difference methods to generate the formulas.

If the differential equation $y' = f(x, y)$ is integrated between the limits x_n and x_{n+k}, the result will be

$$y(x_{n+k}) - y(x_n) = \int_{x_n}^{x_{n+k}} f(x, y)\, dx$$

This integral cannot be found directly since the relationship $y'(x)$ is not known in advance. A number of finite-difference methods to approximate this integral do exist. The particular selection will characterize the individual method being used. Any numerical integration formula that does not require a prior estimate of $y'(x_{n+1})$ is suitable for use as a predictor. Any numerical integration formula that requires a prior estimate of $y'(x_{n+1})$ is suitable for use as a corrector.

Milne's Predictor-Corrector Method

Milne's method [12] uses Milne's formula,

$$y_{n+1} = y_{n-3} + \frac{4h}{3} [2y'_n - y'_{n-1} + 2y'_{n-2}] + \frac{28}{90} h^5 y^{(5)}$$

as a predictor and Simpson's rule,

$$y_{n+1} = y_{n-1} + \frac{h}{3} [y'_{n+1} + y'_n + y'_{n-1}] - \frac{1}{90} h^5 y^{(5)}$$

as a corrector. The final terms in each of these formulas are not actually used in the iteration process. Rather, they are included because they give an indication of the truncation error. Milne's method is termed a fourth-order method since the truncation error is on the order of h^5. The user may be tempted to ask why it is necessary to bother with the corrector at all if the predictor is of fourth order. The answer to this is revealed by the relative size of the two error terms. In this case the corrector truncation error is 28 times smaller and is thus much more desirable. In general, iterative formulas are substantially more accurate than forward formulas. For this reason, they are worthwhile to use in spite of their added difficulty. Although Milne's formulation has a small numerical coefficient in its error remainder ($\frac{1}{90}$), it is less often used than other methods with poorer remainders owing to the fact that it is inherently unstable. This means that the propagated error may grow exponentially. This characteristic is true for all corrector formulas based on Simpson's rule.

Adams-Bashforth Method

The Adams–Bashforth method [12] is a fourth-order method that has its predictor equation based on an integration of Newton's backward interpolation formula. For this method, the predictor formula is

$$y_{n+1} = y_n + \frac{h}{24} [55y'_n - 59y'_{n-1} + 37y'_{n-2} - 9y'_{n-3}] + \frac{251}{720} h^5 y^{(5)}$$

and the corrector formula is

$$y_{n+1} = y_n + \frac{h}{24}[9y'_{n+1} + 19y'_n - 5y'_{n-1} + y'_{n-2}] - \frac{19}{720}h^5y^{(5)}$$

This method proceeds just as does Milne's method, but an error introduced at one stage in the Adams–Bashforth procedure does not tend to grow exponentially.

One might reasonably suspect that, since the estimate of the error term is known, it could be used to improve the corrected value. Although this could be done, Ralston [19] comments that this process is equivalent to using a system of one higher order. Since the process of correcting the corrector can adversely influence the stability of the corrector, it is a better idea to use a higher-order formulation if higher accuracy is desired.

Hamming's Method

Hamming's method [12] is based on the following computational formulas:

Predictor: $$y_{n+1}^{(0)} = y_{n-3} + \frac{4h}{3}(2y'_n - y'_{n-1} + 2y'_{n-2})$$

Modifier: $$\bar{y}_{n+1}^{(0)} = y_{n+1}^{(0)} + \frac{112}{121}[y_n - y_n^{(0)}]$$

$$[\bar{y}_{n+1}^{(0)}]' = f[x_{n+1}, \bar{y}_{n+1}^{(0)}]$$

Corrector: $$y_{n+1}^{(j+1)} = \frac{1}{8}(9y_n - y_{n-2}) + \frac{3h}{8}\{[y_{n+1}^{(j)}]' + 2y'_n - y'_{n-1}\}$$

This stable, fourth-order method is based on the predictor equation

$$y_{n+1} = y_{n-3} + \frac{4h}{3}[2y'_n - y'_{n-1} + 2y'_{n-2}] + \frac{28}{90}h^5y^{(5)}$$

and a corrector equation

$$y_{n+1} = \tfrac{1}{8}[9y_n - y_{n-2} + 3h(y'_{n+1} + 2y'_n - y'_{n-1})] - \tfrac{1}{40}h^5y^{(5)}$$

The method has an added feature that uses estimates of the errors in the predictor and corrector to "mop up" their errors. Of all the predictor–corrector methods, Hamming's method is one of the most often used methods because of its simplicity and stability.

5.4 SUMMARY OF PREDICTOR-CORRECTOR CHARACTERISTICS

When compared with one-step methods, predictor–corrector methods have certain important characteristics.

1. Predictor–corrector methods require information about prior points and therefore are not self-starting. Indeed, they must rely on some type of one-step methods to get their start. If a change in step size is made during the solution process, a temporary reversion to the one-step starter method is usually required.
2. Since the predictor–corrector methods need information about prior points, they also require the computer capacity to store this information.
3. The one-step methods are of comparable accuracy to the predictor–corrector methods of the same order. The predictor–corrector methods provide an easy measure of the per step error, while the one step methods do not. For this reason, the step size h is often chosen conservatively smaller for the one-step methods than would otherwise be necessary. This tends to make the predictor–corrector methods appear to be more efficient.
4. The actual number of functional evaluations for each step of a fourth-order Runge–Kutta method will be four, whereas a predictor–corrector method of the same order will often require only two evaluations to achieve convergence. For this reason the predictor–corrector methods can be almost twice as fast as the Runge–Kutta methods of comparable accuracy. This time saving can become a significant consideration in the selection of an algorithm for the microcomputer.

5.5 STEP-SIZE CONSIDERATIONS

One significant practical problem faced by the scientific programmer who is solving a differential equation on a microcomputer is that of selecting a suitable step size for the computational process. If the step size is too small, the computational process will consume needless computer time and the number of per step errors contributing to the global error will be significant. If, on the other hand, the step size is too large, the local error due to truncation will be significant, and the resulting accumulation of global errors will cause the computational results to be of poor accuracy.

The most common procedure used in selecting the step size is to keep the local per step error below a predetermined, allowable value.

In general, for a method of order n the local error will be on the order of a constant times the step size raised to the power $n + 1$. This may be expressed as

$$Ch^{n+1}$$

If the method being used is a predictor–corrector method, the per step error is often presented as the last term in the corrector formula (see for example the discussion of Milne's method). If the Runge–Kutta method is being used, however, the local error is not so obvious. One way to estimate this error is based on Richardson [20] extrapolation and will now be discussed.

If the step size h is used to predict a value y_{j+1} of the solution at the point x_{j+1}, then the difference between this value and the true value y_{true} will be

$$y_{\text{true}} - y_{j+1} = Ch^{n+1}$$

If a step size of $h/2$ is used to predict the value y_{j+1}^* at x_{j+1}, then the difference between this new value and the true value will be

$$y_{\text{true}} - y_{j+1}^* = C\left(\frac{h}{2}\right)^{n+1}$$

If this equation is subtracted from the previous expression, the result will be

$$\begin{aligned} y_{j+1} - y_{j+1}^* &= Ch^{n+1} - C\left(\frac{h}{2}\right)^{n+1} \\ &= Ch^{n+1}\left(1 - \left(\frac{1}{2}\right)^{n+1}\right) \end{aligned}$$

Thus one can solve for the estimate of the local error:

$$Ch^{n+1} = \frac{(y_{j+1} - y_{j+1}^*)(2^{n+1})}{2^{n+1} - 1}$$

The disadvantage to this method is that it requires the user to compute the value of y_{j+1} twice. Since the y_{j+1}^* computation requires the two half-size steps to get the value at x_{j+1}, the computational effort is more than doubled. Nevertheless, this procedure is often included in the computational algorithm to make automatic adjustments in the step size as the computational process takes place. This approach is frequently used for Runge–Kutta routines. Alternatively, if the per step error is too large for a given step size, the error could be reduced by utilizing a higher-order term in the computational process. This is, of course, most easily accomplished for the predictor–corrector methods.

The chief advantage of the Runge–Kutta methods is their ability to start easily and the ease with which the step sizes can be changed during the computational process. On the other hand, the primary advantage of the predictor–corrector methods is the ease with which the per step error can be estimated. Although in the past these advantages were often regarded as mutually exclusive, recent powerful techniques allow the user to exploit the computational advantages of both types of methods. Such hybrid methods can be quite useful in the solution of engineering problems. As an example, Gear [8] presents a method that features automatic control of both the computational step size and the degree of the predictor–corrector formulas.

5.6 STIFF PROBLEMS

Some types of ordinary differential equation problems do not lend themselves to solution by any of the methods previously discussed. To understand why this is so requires an understanding of the component parts to the solution of a differential equation. The time constant of a first-order differential equation is the time required for the transient portion of the solution to decay by a factor of $1/e$. A differential equation of nth order will in general have n time constants. If any two of these time constants are widely different in magnitude or if one of the time constants is quite small relative to the solution interval, the problem is said to be *stiff* and will perform poorly when treated by traditional methods. Such systems require that the solution technique have small enough step sizes to account for the fastest component part of the process even after the contribution of this part has died out. Failure to maintain a small enough step size will lead to instability in the solution process. Although the difficulty in maintaining stability when using traditional methods on stiff systems can be temporarily averted by using small step sizes, this approach has two disadvantages. First, the use of extremely small step sizes relative to the size of the solution interval will cause the method to consume considerable time in achieving a solution. Second, the round-off and truncation errors that are amplified and accumulated through the use of many calculations will eventually lead to meaningless results.

Since stiff problems can occur in process control problems, electronic network problems, and chemical reaction problems, a number of recent research efforts have been directed toward discovering efficient methods for treating such problems. Although a discussion of these methods is beyond the scope of this text, the interested reader is encouraged to consult the works of Gear [7] or Hall and Watt [11].

5.7 METHODS FOR THE SOLUTION OF BOUNDARY-VALUE PROBLEMS

As mentioned previously, any ordinary differential equation that has constraints imposed at values of the independent variable other than at zero is called a boundary-value problem. If only one condition were specified in a boundary-value problem (as would be the case for a first-order differential equation), the independent variable could be adjusted by a change of variable to transform the boundary-value problem into an initial-value problem. For this reason, it only makes sense to deal with boundary-value problems that are of second order or higher.

For simplicity in this discussion, we will speak in terms of a second-order equation:

$$\frac{d^2y}{dx^2} = f(x, y, y')$$

with boundary conditions

$$y(a) = A \quad \text{and} \quad y(b) = B$$

even though equations of higher order can be treated by the same techniques. Solution methods for this system fall into two general categories:

1. Techniques that rely on reducing the problem to that of solving multiple initial-value problems.
2. Techniques that employ a finite-difference form of the differential equation.

Initial-Value Methods (Shooting Methods)

If the second-order differential equation is linear of the form

$$y'' = f_1(x)y' + f_2(x)y + f_3(x)$$
$$y(a) = A, \qquad y(b) = B$$

it may be reduced to an initial-value problem by means of the initial conditions:

$$y(a) = A \quad \text{and} \quad y'(a) = \alpha_1$$

Once a solution $y_1(x)$ is known, a different set of boundary conditions

$$y(a) = A \quad \text{and} \quad y'(a) = \alpha_2$$

may be applied to achieve a second solution, $y_2(x)$. If $y_1(b) = \beta_1$ and $y_2(b) = \beta_2$, then if $\beta_1 \neq \beta_2$, the solution

$$y(x) = \frac{1}{\beta_1 - \beta_2} [(B - \beta_2)\, y_1(x) + (\beta_1 - B)\, y_2(x)]$$

satisfies both of the original boundary conditions.

If the differential equation is nonlinear, a series of initial-value problems may be solved using successive values of α in the initial conditions

$$y(a) = A \quad \text{and} \quad y'(a) = \alpha$$

in an attempt to find a solution that will ultimately satisfy $y(b) = B$. In such a process, interpolation can often be used to suggest an orderly progression for α that will minimize wasted effort. Unfortunately, this technique is inherently inefficient and is not recommended as a substitute for some of the advanced methods available in the literature.

Finite-Difference Methods

The advantage to using a finite-difference approach is that the solution of the boundary-value problem may be reduced to that of solving a system of algebraic equations. To treat the two-point boundary-value problem

$$y'' = f(x, y, y')$$

$$y(a) = A, \qquad y(b) = B$$

the interval from a to b may be divided into n equal parts

$$x_i = x_0 + ih, \qquad i = 1, 2, \ldots, n$$

where $x_0 = a$, $x_n = b$, and

$$h = \frac{b - a}{n}$$

The x_i points are called *node* points and represent locations at which the ordinate values y_i of the solution are desired. Using these node locations and finite-difference relationships for the derivatives

$$y'(x_i) = \frac{1}{2h}(y_{i+1} - y_{i-1})$$

$$y''(x_i) = \frac{1}{h^2}(y_{i+1} - 2y_i + y_{i-1})$$

the differential equation can be written as a *difference equation*. (It should be noted that various types of finite-difference forms exist for the expression of derivatives. The topic of difference relationships will be discussed in more depth in Chapter 7.) If this difference equation is written for $i = 1, 2, \ldots, n$ and the two boundary conditions are used, the problem is reduced to a system of $n - 1$ equations in $n - 1$ algebraic unknowns y_i. If the original differential equation is linear, the problem will give rise to the simultaneous solution of a set of linear algebraic equations. If, on the other hand, the original differential equation is nonlinear, the problem solution will become that of solving a system of simultaneous, nonlinear algebraic equations. Although the solution of algebraic equations is well understood, the solution of boundary-value problems by the method of finite differences is difficult to systematize into a "prepackaged" computer subroutine, since the formulation of each problem depends on the nature of the specific differential equation being considered.

EXAMPLE 5-3 Suppose that it is desired to solve the differential equation

$$y'' = 2x + 3y$$

$$y(0) = 0, \qquad y(1) = 1$$

using $h = 0.2$. The differential equation can be written in finite-difference form as

$$\frac{1}{0.04}[y_{i+1} - 2y_i + y_{i-1}] = 2x_i + 3y_i$$

This formula and the boundary conditions can be used to write the following system of four linear equations in four unknowns:

$$-2.12y_1 + y_2 = 0.016$$
$$y_3 - 2.12y_2 + y_1 = 0.032$$
$$y_4 - 2.12y_3 + y_2 = -0.064$$
$$-2.12y_4 + y_3 = -0.936$$

The following table represents a comparison of the results of solving this system and the exact solution, which is

$$y(x) = \frac{5 \sinh \sqrt{3x}}{3 \sinh \sqrt{3}} - \frac{2}{3}$$

x	y (*numerical solution*)	y (*exact*)
0.0	0.0	0.0
0.2	0.0827	0.0818
0.4	0.1912	0.1897
0.6	0.3548	0.3529
0.8	0.6088	0.6073
1.0	1.0000	1.0000

5.8 CONSIDERATIONS IN THE SELECTION OF AN ALGORITHM FOR SOLVING ORDINARY DIFFERENTIAL EQUATIONS

Although it is impossible to state universal rules that will guide the user in selecting a best-method microcomputer algorithm for solving a given ordinary differential equation, a few basic guidelines do exist. These are listed in this section.

1. **Consider the type of problem.** If the problem to be solved is an initial-value problem, many different types of systematic algorithms are available to implement the solution. If the problem is a boundary-value problem, the user may have to write his or her own special-purpose software.

2. **Consider the complexity of the differential equation.** If an initial-value problem is quite complicated and the evaluation of $f(x, y)$ is rather involved, a predictor–corrector method is usually preferred since it requires only two evaluations of $f(x, y)$ for each step, rather than the four required for Runge–Kutta methods. For multipurpose uses where the evaluation of $f(x, y)$ is not difficult, the Runge–Kutta method is often more convenient.
3. **Consider the time involved in solving the problem.** If computer time cost is a premium factor, the best method to use may be a predictor–corrector method. If, on the other hand, the user preparation time is the critical factor, a method such as the Runge–Kutta method may be preferred. There will also be a tradeoff between run time and accuracy of the result, since larger step sizes usually give faster results but lower accuracy.
4. **Consider the accuracy required.** If the accuracy of the required result of a differential equation to be solved is known in advance of the solution, a method with adjustment of the step size similar to that implemented in Example 5-2 is desirable.

REFERENCES

1. BENNETT, A.W., *Introduction to Computer Simulation*, West Publishing Co., New York, 1974.
2. CESHINO, F., and J. KUNTZMANN, *Numerical Solution of Initial Value Problems*, Prentice-Hall, Inc., Englewood Cliffs, N.J., 1966.
3. DAHLQUIST, G., and A. BJORCK, *Numerical Methods*, Prentice-Hall, Inc., Englewood Cliffs, N.J., 1974.
4. DORN, W.S., and D.D. McCRACKEN, *Numerical Methods with FORTRAN IV Case Studies*, John Wiley & Sons, Inc., New York, 1972.
5. FORSYTHE, G.E., M.A. MALCOLM, and C.B. MOLER, *Computer Methods for Mathematical Computations*, Prentice-Hall, Inc., Englewood Cliffs, N.J., 1977.
6. GEAR, C.W., *Numerical Initial Value Problems in Ordinary Differential Equations*, Prentice-Hall, Inc., Englewood Cliffs, N.J., 1971.
7. GEAR, C.W., "Hybrid Methods for Initial Value Problems in Ordinary Differential Equations," *SIAM Journal of Numerical Analysis*, Ser. B, vol. 2, no. 1, 1964, 69–86.
8. GEAR, C.W., and D.S. WATANABE, "Stability and Convergence of Variable Order Multistep Methods," *SIAM Journal of Numerical Analysis*, vol. 11, no. 5, 1974, 1044–1058.
9. GEAR, C.W., and K.W. TU, "The Effect of Variable Mesh Size on the Stabil-

ity of Multistep Methods," *SIAM Journal of Numerical Analysis*, vol. 11, no. 5, 1974, 1025–1043.

10. GROVE, W.E., *Brief Numerical Methods*, Prentice-Hall, Inc., Englewood Cliffs, N.J., 1966.
11. HALL, G., and J.M. WATT, *Modern Numerical Methods for Ordinary Differential Equations*, Oxford University Press, Inc., New York, 1976.
12. HAMMING, R.W., "Stable Predictor–Corrector Methods for Ordinary Differential Equations," *Journal of the Association of Computing Machinery*, vol. 6, 1959, 37–47.
13. JENNINGS, W., *First Course in Numerical Methods*, Macmillan, Inc., New York, 1964.
14. KOCHENBURGER, R.J., *Computer Simulation of Dynamic Systems*, Prentice-Hall, Inc., Englewood Cliffs, N.J., 1972.
15. LAPIDUS, L., and J.H. SEINFELD, *Numerical Solution of Ordinary Differential Equations*, Academic Press, Inc., New York, 1971.
16. McCALLA, T.R., *Introduction to Numerical Methods and FORTRAN Programming*, John Wiley & Sons, Inc., New York, 1967.
17. PALL, G.A., *Introduction to Scientific Computing*, Appleton-Century-Crofts Educational Division, Meredith Corp., New York, 1971.
18. RALSTON, A., *A First Course in Numerical Analysis*, McGraw-Hill Book Co., New York, 1965.
19. RALSTON, A., and H.S. WILF, *Mathematical Methods for Digital Computers*, John Wiley & Sons, Inc., New York, 1967.
20. RICHARDSON, L.F., and J.A. GAUNT, "The Deferred Approach to the Limit," *Transactions of the Royal Society*, *London*, vol. 226A, 1927, 300.
21. WILLIAMS, P.W., *Numerical Computation*, Barnes & Noble Books, New York, 1972.

The microcomputer frequently forms the nerve center for a complete system for solving special engineering and scientific tasks. (*Photo courtesy of Bruning.*)

Numerical interpolation and curve fitting

6

The process of scientific investigation frequently provides considerable quantitative output. It is this quantitative nature of scientific and technological problems that leads to the processing challenges of numerical information. Indeed, the key to understanding for many problems is often revealed as a result of a thoughtful presentation of the numerical data describing the problem. Conversely, poor data presentation can often lead to confusion and misinterpretation and can increase the opportunity for human error. Because of its portability and numerical capabilities, the microcomputer is frequently used in the laboratory setting for the analysis of test data. In the manipulation and management of scientific and engineering data, certain basic computational tools have become extremely useful. Among these are the following:

1. Numerical interpolation
2. Curve fitting
3. Numerical differentiation
4. Numerical integration

It is the purpose of this chapter to discuss the first two of these topics within the context of engineering data management. The topics of differentiation and integration will be treated in Chapter 7.

6.1 LINEAR INTERPOLATION

Engineering and scientific data are frequently available in tabular form. This method of data representation arises owing to the fact that the data were obtained only for discrete values by experimental means or owing to the fact that practical limitations on the volume of data to be managed requires that only a few values be stored. The process of interpolation is that of finding a value of a function at some intermediate location in the tabular data.

The simplest form of interpolation is linear interpolation based on a linear approximation through points (x_k, y_k) and (x_{k+1}, y_{k+1}), as shown in Figure 6-1. The equation of the line is

$$\frac{y - y_k}{x - x_k} = \frac{y_{k+1} - y_k}{x_{k+1} - x_k}$$

or

$$y = \frac{y_k(x - x_{k+1}) - y_{k+1}(x - x_k)}{x_k - x_{k+1}}$$

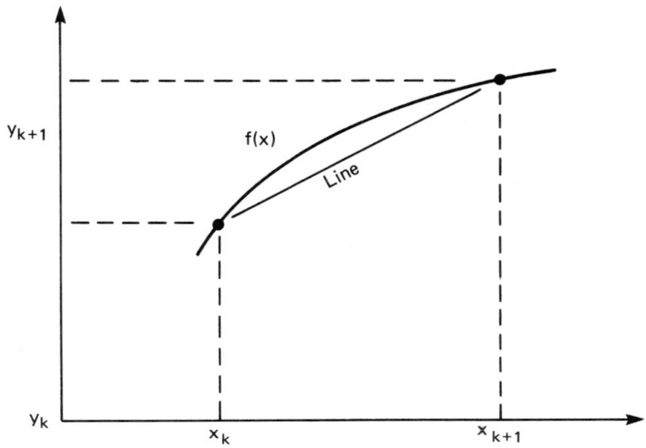

Figure 6-1 Linear interpolation.

Thus if one knows two tabular values that surround a given x value, this linear relationship can be used to find an approximation to the corresponding y value. It is generally accepted that the use of more neighboring points and an approximation of more complexity than a line will give better results. In the following sections, methods will be developed to find a unique nth-order polynomial $P_n(x)$ that approximates the function $f(x)$ by satisfying all $n + 1$ tabular points (x_i, y_i), $i = 0, 1, \ldots, n$. The polynomial is said to have constraints

$$P_n(x_i) = y_i, \qquad i = 0, \ldots, n$$

The methods of finding this polynomial fall into three categories: Lagrangian methods, difference methods, and iterative methods. These three will now be discussed.

6.2 LAGRANGE INTERPOLATION

In Lagrangian interpolation, $n + 1$ table values (x_i, y_i), $i = 0, \ldots, n$, are given representing points on the function $y = f(x)$ on the interval $x_0 \leqslant x \leqslant x_n$. For this method the interpolating polynomial will be expressed as

$$P_n(x) = y_0 b_0(x) + y_1 b_1(x) + \cdots + y_n b_n(x)$$

where each $b_j(x)$ is a polynomial of degree n. These polynomials can be determined by imposing the $n + 1$ constraint equations

$$P_n(x_i) = y_i, \qquad i = 0, \ldots, n$$

This will give a system of the form

$$\begin{aligned} &y_0 b_0(x_0) + y_1 b_1(x_0) + \cdots y_n b_n(x_0) = y_0 \\ &\vdots \\ &y_0 b_0(x_n) + \qquad \cdots \qquad y_n b_n(x_n) = y_n \end{aligned}$$

If the $b_j(x_i)$ values are chosen so that

$$b_j(x_i) = \begin{cases} 1, & i = j \\ 0, & i \neq j \end{cases}$$

then the constraint equations will be satisfied. This condition requires that each $b_j(x)$ have zeros corresponding to every value of x except x_j. A general $b_j(x)$ polynomial that always satisfies this condition is

$$b_j(x) = c_j(x - x_0)(x - x_1) \cdots (x - x_{j-1})(x - x_{j+1}) \cdots (x - x_n)$$

Notice that this polynomial does not have a term of the form $(x - x_j)$. Since it is required that

$$b_j(x_j) = 1$$

the coefficient c_j can be found to be

$$c_j = \frac{1}{(x_j - x_0)(x_j - x_1) \cdots (x_j - x_{j-1})(x_j - x_{j+1}) \cdots (x_j - x_n)}$$

Thus the final polynomial will be

$$P_n(x) = \sum_{j=0}^{n} y_j \frac{(x - x_0)(x - x_1) \cdots (x - x_{j-1})(x - x_{j+1}) \cdots (x - x_n)}{(x_j - x_0)(x_j - x_1) \cdots (x_j - x_{j-1})(x_j - x_{j+1}) \cdots (x_j - x_n)}$$

This polynomial can be written in an easier form using

$$Z_j(x) = (x - x_0)(x - x_1) \cdots (x - x_{j-1})(x - x_{j+1}) \cdots (x - x_n)$$

Thus

$$P_n(x) = \sum_{j=0}^{n} y_j \frac{Z_j(x)}{Z_j(x_j)}$$

To illustrate the use of Lagrange interpolation, a numerical application will now be presented.

EXAMPLE 6-1 Suppose that a table of information is given:

x_i	y_i
10.	0.17365
20.	0.34202
30.	0.50000
40.	0.64279
50.	0.76604
60.	0.86603

These data correspond to the function $y = \sin(x_{\text{degrees}})$. Find the value of y at $x = 23$ using the method of Lagrange interpolation.

In applying the method of Lagrange interpolation to these data, the numerical value of $P(23)$ is assembled using the summation formula given at the end of Section 6.2. The simplicity of this process is obvious from the computer program listed next. The actual Lagrange interpolation step is accomplished using two FOR loops in statements 3340 through 3440 of this BASIC program.

```
1000  REM **********************
1010  REM *THIS PROGRAM FINDS  *
1020  REM *AN INTERMEDIATE     *
1030  REM *VALUE OF A TABLE OF *
1040  REM *VALUES BY THE METHOD*
1050  REM *OF LAGRANGE         *
1060  REM *INTERPOLATION       *
1090  REM **********************
1100 :
1110 :
1120  REM **SET UP THE TABLE
1130 :
1135 M = 5
1140  DIM X(6),Y(6)
1150  PRINT "--------------------
      "
1155  PRINT " X            Y
      "
1157  PRINT "--------------------
      "
1160  FOR I = 0 TO M
1180  READ X(I),Y(I)
1185  PRINT X(I); TAB( 10);Y(I)
1190  NEXT I
1210  DATA  10.,0.17365,20.,.3420
      2
1220  DATA  30.,0.50000,40.,.6427
      9
1230  DATA  50.,0.76604,60.,.8660
      3
1240  PRINT "--------------------
      "
```

```
1250 :
1260 XV = 23.
1270  REM **FIND THE VALUE**
1290 :
1300  GOSUB 3000
1310 :
1320 :
1330  REM **WRITE THE ANSWER**
1340 :
1420  PRINT "THE VALUE AT X=";XV
1425  PRINT "IS Y=";YV
1430 :
1440  END
1450 :
1460 :
1470 :
3000  REM *********************
3010  REM * THIS SUBROUTINE    *
3020  REM * APPLIES LAGRANGE   *
3030  REM * INTERPOLATION TO   *
3040  REM * GET AN INTERMEDIATE*
3050  REM * VALUE FROM A TABLE *
3060  REM * OF VALUES.         *
3070  REM *                    *
3080  REM *   PARAMETERS:      *
3090  REM *                    *
3100  REM *   X(I)  - ARRAY OF *
3110  REM *           ARGUMENT *
3120  REM *           VALUES OF*
3130  REM *           THE TABLE*
3140  REM *                    *
3150  REM *   Y(I)  - ARRAY OF *
3160  REM *           FUNCTION *
3170  REM *           VALUES OF*
3180  REM *           THE TABLE*
3190  REM *                    *
3200  REM *   M     - DIMENSION*
3210  REM *           OF TABLE *
3220  REM *                    *
3230  REM *   XV    - INPUT    *
3240  REM *           ARGUMENT *
3250  REM *                    *
3260  REM *   YV    - OUTPUT   *
3270  REM *           VALUE    *
3280  REM *           FOUND    *
3290  REM *                    *
3300  REM *********************
3310 :
3320 :
3330 YV = 0.
3340  FOR J = 0 TO M
3350 ZN = 1.:ZD = 1.
3360 :
3370  FOR I = 0 TO M
3380  IF (I = J) GOTO 3410
3390 ZN = ZN * (XV - X(I))
3400 ZD = ZD * (X(J) - X(I))
3410  NEXT I
3420 :
3430 YV = YV + Y(J) * ZN / ZD
3440  NEXT J
3450 :
3460  RETURN
```

The output of this program is as follows:

```
--------------------
 X          Y
--------------------
10        .17365
20        .34202
30        .5
40        .64279
50        .76604
60        .86603
--------------------
THE VALUE AT X=23
IS Y=.390730172
```

To reach this output, an Apple II computer required less than 2 seconds. The primary advantage to Lagrange interpolation is its speed and simplicity. The disadvantage to the method is that the complete program must be rerun in order to find the interpolated value for some other x value in the table.

6.3 METHOD OF DIVIDED DIFFERENCES

The Lagrange interpolation method works well if one interpolated value is all that is required. For some types of scientific and engineering problems, many intermediate values must be interpolated from a given table of data values. When this happens, it is desirable to have an interpolation method that is more efficient of computational time and effort. One type of method that exhibits this advantage is the method of divided differences. This family of methods requires a bit more data manipulation in order to set up the interpolation scheme, but it does allow the user to obtain a number of interpolated values without the need to go through the initial setup of the problem.

Although there are a number of difference based formulations for interpolation, the Newton's forward difference formulation, also known as the Newton Gregory formulation, is the most commonly used. It is based on an interpolating polynomial of the form

$$P_n(x) = c_0 + c_1(x - x_0) + c_2(x - x_0)(x - x_1) + \cdots + c_n(x - x_0)(x - x_1)\cdots(x - x_{n-1})$$

The coefficients c_j can be found by applying the constraint equations

$$P_n(x) = y_i, \qquad i = 0, \ldots, n$$

This would give a system of equations of the form

$$\begin{aligned} c_0 &= y_0 \\ c_0 + c_1(x_1 - x_0) &= y_1 \\ c_0 + c_1(x_2 - x_0) + c_2(x_2 - x_0)(x_2 - x_1) &= y_2 \\ &\vdots \\ c_0 \cdots + c_n(x_n - x_0)(x_n - x_1)\cdots(x_n - x_{n-1}) &= y_n \end{aligned}$$

This form is fortunately linear and triangular. The values of c_j can be found without difficulty. If the data values are equally spaced, there is an easier way to find them based on forward finite differences. If the x values are equally spaced, then

$$x_{i+1} - x_i = h$$

Then, in general, $x_i = x_0 + ih$ for $i = 1, \ldots, n$. When this expression is used, the equations to be solved are

$$
\begin{aligned}
y_0 &= c_0 \\
y_1 &= c_0 + c_1 h \\
y_2 &= c_0 + c_1(2h) + 2h^2 c_2 \\
&\vdots \\
y_i &= c_0 + c_1 ih + c_2 ih[(i-1)h] + \cdots + c_i(i!)\,h^i
\end{aligned}
$$

If one solves for the coefficients, the result will be

$$
\begin{aligned}
c_0 &= y_0 \\
c_1 &= \frac{y_1 - c_0}{h} = \frac{y_1 - y_0}{h} = \frac{\Delta y_0}{h}
\end{aligned}
$$

In this expression, Δy_0 is called the first forward difference. As the process continues,

$$
\begin{aligned}
c_2 &= \frac{1}{2h^2}[y_2 - c_0 - 2hc_1] = \frac{1}{2h^2}[(y_2 - y_1) - (y_1 - y_0)] \\
&= \frac{1}{2h^2}[\Delta(\Delta y_0)] = \frac{\Delta^2 y_0}{2h^2}
\end{aligned}
$$

In this expression, $\Delta^2 y_0$ is called the second forward difference since it is the difference of the differences. In general, the c_j coefficients of the polynomial can be expressed as

$$
c_j = \frac{\Delta^j y_0}{(j!)\,h^j}
$$

In general, the higher-order differences of the function $y = f(x)$ are defined over the interval $x_0 \leqslant x \leqslant x_n$ as

$$
\Delta^j y_i = \Delta^{j-1} y_{i+1} - \Delta^{j-1} y_i, \qquad i = 0, \ldots, n-j
$$

Frequently, these differences are tabulated as shown in Table 6-1. In this table the differences of any given order depend on the differences of the next lower order. To illustrate the use of the Newton forward difference interpolation method, the previous example application will be reworked.

Table 6-1 Forward Differences

x_i	y_i	$\Delta y_i = y_{i+1} - y_i$	$\Delta^2 y_i = \Delta y_{i+1} - \Delta y_i$	$\Delta^3 y_i = \Delta^2 y_{i+1} - \Delta^2 y_i$	$\Delta^4 y_i = \Delta^3 y_{i+1} - \Delta^3 y_i$	$\Delta^5 y_i = \Delta^4 y_{i+1} - \Delta^4 y_i$
x_0	y_0					
		Δy_0				
x_1	y_1		$\Delta^2 y_0$			
		Δy_1		$\Delta^3 y_0$		
x_2	y_2		$\Delta^2 y_1$		$\Delta^4 y_0$	
		Δy_2		$\Delta^3 y_1$		$\Delta^5 y_0$
x_3	y_3		$\Delta^2 y_2$		$\Delta^4 y_1$	
		Δy_3		$\Delta^3 y_2$		$\vdots$
x_4	y_4		$\Delta^2 y_3$		$\vdots$	
		Δy_4		$\vdots$		
x_5	y_5		$\vdots$			
		$\vdots$				
$\vdots$	$\vdots$					

EXAMPLE 6-2 Use the method of divided differences to find the value of $y(23)$ using the data from Example 6-1.

x_i	y_i
10.	0.17365
20.	0.34202
30.	0.50000
40.	0.64279
50.	0.76604
60.	0.86603

Using these data, a difference table can be constructed as follows:

x_i	y_i	Δy_i	$\Delta^2 y_i$	$\Delta^3 y_i$	$\Delta^4 y_i$	$\Delta^5 y_i$
10.	0.17365	–				
		0.16837	–			
20.	0.34202		−0.01039	–		
		0.15798		−0.00480	–	
30.	0.50000		**−0.01519**		+0.00045	–
		0.14279		**−0.00435**		+0.00018
40.	0.64279		−0.01954		**+0.00063**	
		0.12325		−0.00372		
50.	0.76604		−0.02326			
		0.09999				
60.	0.86603					

The value of x_0 may be chosen anywhere in the table. Suppose that the value $x_0 = 20$ is chosen. The necessary differences lie on a diagonal down from x_0. As many or as few higher-order differences may be used as desired. In general, the accuracy will improve if more differences are used. One advantage to this method is that it allows the user to add differences to a previous calculation without the need to start over in the calculation process. Thus, if one does not know how many terms to carry along, terms can be added until the contribution of the added terms is small enough so that the number of decimal digits has stabilized. In this example problem, $h = 10$. Using only first-order differences gives

$$y(23.) = y_0 + \frac{\Delta y_0}{h}(23. - x_0) = 0.34202 + \frac{0.15798}{10}(3)$$

$$= 0.38941$$

Using first- and second-order differences gives

$$y(23.) = 0.38941 + \frac{\Delta^2 y_0}{2h^2}\,(23 - x_0)(23 - x_1)$$

$$= 0.38941 + \frac{-0.01519(3)(-7)}{200.} = 0.39100$$

Using first-, second-, and third-order differences gives

$$y(23.) = 0.39101 + \frac{\Delta^3 y_0}{6h^3}\,(23 - x_0)(23 - x_1)(23 - x_2)$$

$$= 0.39074$$

Clearly, this answer approaches the actual value 0.39073. Once the table of differences is established, it could be used to find other interpolated values without the need to go through the setup process.

This procedure can be implemented by a microcomputer. A sample BASIC program to perform this task now follows:

```
1000  REM **********************
1010  REM *THIS PROGRAM FINDS   *
1020  REM *AN INTERMEDIATE      *
1030  REM *VALUE OF A TABLE OF  *
1040  REM *VALUES BY THE METHOD*
1050  REM *OF NEWTON FORWARD    *
1060  REM *DIFFERENCE INTERPOL-*
1070  REM *ATION.               *
1080  REM **********************
1090 :
1100 :
1110  REM **SET UP THE TABLE
1120 :
1130 M = 5:H = 10.
1140  DIM X(6),Y(6),D(6)
1150  PRINT "--------------------"
1160  PRINT " X           Y       "
1170  PRINT "--------------------"
1180  FOR I = 0 TO M
1190  READ X(I),Y(I)
1200  PRINT X(I); TAB( 10);Y(I)
1210  NEXT I
1220  DATA  10.,0.17365,20.,.34202
1230  DATA  30.,0.50000,40.,.64279
1240  DATA  50.,0.76604,60.,.86603
1250  PRINT "--------------------"
1260 :
1270 XV = 23.
1280  REM **FIND THE VALUE**
1290 :
1300  GOSUB 3000
1310 :
1320 :
1330  REM **WRITE THE ANSWER**
1340 :
1350  PRINT "THE VALUE AT X=";XV
1360  PRINT "IS Y=";YV
1370 :
1380  END
1390 :
1400 :
1410 :
3000  REM **********************
3010  REM * THIS SUBROUTINE     *
3020  REM * APPLIES NEWTON      *
3030  REM * FORWARD DIFFERENCE  *
3040  REM * INTERPOLATION TO    *
3050  REM * GET AN INTERMEDIATE*
3060  REM * VALUE FROM A TABLE  *
3070  REM * OF VALUES.          *
3080  REM *                     *
```

```
3090  REM *                       *
3100  REM *   PARAMETERS:         *
3110  REM *                       *
3120  REM *   X(I)  - ARRAY OF    *
3130  REM *           ARGUMENT    *
3140  REM *           VALUES OF*
3150  REM *           THE TABLE*
3160  REM *                       *
3170  REM *   Y(I)  - ARRAY OF    *
3180  REM *           FUNCTION    *
3190  REM *           VALUES OF*
3200  REM *           THE TABLE*
3210  REM *                       *
3220  REM *   M     - DIMENSION*
3230  REM *           OF TABLE    *
3240  REM *                       *
3250  REM *   H     - UNIFORM     *
3260  REM *           SPACING     *
3270  REM *           OF X(I)     *
3280  REM *                       *
3290  REM *   XV    - INPUT       *
3300  REM *           ARGUMENT    *
3310  REM *                       *
3320  REM *   YV    - OUTPUT      *
3330  REM *           VALUE       *
3340  REM *              FOUND    *
3350  REM *                       *
3360  REM ********************
3370 :
3380 :
3390 YV = Y(0):FACT = 1.
3400 :
3410  FOR I = 0 TO M
3420 D(I) = Y(I)
3430  NEXT I
3440 :
3450  FOR OD = 1 TO M
3460  FOR I = 0 TO M - OD
3470 D(I) = D(I + 1) - D(I)
3480  NEXT I
3490 FACT = FACT * OD
3500 C = D(0) / (FACT * H ^ OD)
3510 PR = 1.
3520  FOR J = 0 TO OD - 1
3530 PR = PR * (XV - X(J))
3540  NEXT J
3550 YV = YV + C * PR
3560  PRINT YV
3570  NEXT OD
3580  RETURN
```

The output of this program is as follows:

```
--------------------
 X          Y
--------------------
10        .17365
20        .34202
30        .5
40        .64279
50        .76604
60        .86603
--------------------
.392531
.39050495
.39072335
.390732052
.390730172
THE VALUE AT X=23
IS Y=.390730172
```

To reach this output, an Apple II computer required less than 2 seconds. This program has been designed to print the intermediate interpolated values. The reader will note that the accuracy of these values increases as more terms are added to the interpolation process. This program does not store the difference table, and thus, as it stands, the program could not be used to find other intermediate values without restarting. This added capability would require that additional array storage be used. For some types of computational problems, this tradeoff between storage space and simplicity may well be justified.

Other difference forms can be applied to achieve alternative interpolation schemes. Among these methods are the Newton's backward difference formulation, Gauss's forward difference formulation, and Gauss's backward difference formulation.

6.4 ITERATIVE INTERPOLATION METHODS

Iterative interpolation schemes are based on the repeated application of a simple interpolation process. The best known of these methods is Aitken's method, which is based on repeated linear interpolation. This method will now be described.

It was shown previously that a linear interpolation between points (x_0, y_0) and (x_i, y_i) will be

$$y_{i1}(x) = \frac{1}{x_i - x_0} [y_0(x_i - x) - y_i(x_0 - x)]$$

Using this relationship, a table of values $y_{i1}(x)$, $i = 1, \ldots, n$, can be generated for a given x value. Using these values and linear interpolation of the form

$$y_{i2}(x) = \frac{1}{x_i - x_1} [y_{11}(x)(x_i - x) - y_{i1}(x)(x_1 - x)]$$

a new family of relationships can be found. It can be shown by simple substitution that the $y_{i2}(x)$ relationships are second-order polynomials that satisfy the three points (x_0, y_0), (x_1, y_1), and (x_i, y_i). Once the family of polynomials y_{i2} have been found, linear interpolation using the values of $y_{i2}(x)$ may be accomplished to assemble

$$y_{i3}(x) = \frac{1}{x_i - x_2} [y_{22}(x)(x_i - x) - y_{i2}(x_2 - x)]$$

which is a third-order polynomial that goes through the points (x_0, y_0), (x_1, y_1), (x_2, y_2), and (x_i, y_i). As the process continues, the values of $y_{ij}(x)$ tend to the value of $f(x)$. Although the process can be continued until the limits of the table are reached, to do so may not always guarantee a better answer because of error propagation. It is important to note, however, that Aitken's process does not require uniformly spaced data values. To illustrate this method, Example 6-1 will be interpolated by Aitken's method to find $y(23)$.

EXAMPLE 6-3 Suppose that it is desired to apply Aitken's method to solve the problem posed in Example 6-1. A table of results showing the first three terms of the iterative cycle based on repeated linear interpolation is presented next using $x = 23$. As the method progresses, the table values tend to approach an improved value.

Numerical Results for $x = 23$

i	x_i	y_i	$y_{i1}(23)$	$y_{i2}(23)$	$y_{i3}(23)$
0	10	0.17365	–	–	–
1	20	0.34202	0.39253	–	–
2	30	0.50000	0.38578	0.39051	–
3	40	0.64279	0.37694	0.39019	0.39073
4	50	0.76604	0.36618	0.38990	0.39072
5	60	0.86603	0.35367	0.38962	0.39072

This procedure can be implemented by a microcomputer. The following BASIC program illustrates the use of Aitken's method to solve this interpolation problem.

```
1000  REM *********************
1010  REM *THIS PROGRAM FINDS  *
1020  REM *AN INTERMEDIATE     *
1030  REM *VALUE  OF  A TABLE  *
1040  REM *BY AITKEN'S METHOD  *
1090  REM *********************
1100 :
1110 :
1120  REM **SET UP THE TABLE
1130 :
1135 M = 5
1140  DIM X(6),Y(6),YY(6)
1150  PRINT "-------------------
      "
1155  PRINT " X              Y
      "
1157  PRINT "-------------------
      "
1160  FOR I = 0 TO M
1180  READ X(I),Y(I)
1185  PRINT X(I); TAB( 10);Y(I)
1190  NEXT I
1210  DATA  10.,0.17365,20.,.3420
      2
1220  DATA  30.,0.50000,40.,.6427
      9
1230  DATA  50.,0.76604,60.,.8660
      3
1240  PRINT "-------------------
      "
1250 :
1260 XV = 23.
1270  REM **FIND THE VALUE**
1290 :
1300  GOSUB 3000
1310 :
1320 :
1330  REM **WRITE THE ANSWER**
1340 :
1420  PRINT "THE VALUE AT X=";XV
1425  PRINT "IS Y=";YV
1430 :
1440  END
1450 :
1460 :
1470 :
3000  REM *********************
3010  REM * THIS SUBROUTINE    *
3020  REM * APPLIES  AITKEN'S  *
3030  REM * INTERPOLATION      *
3040  REM * METHOD  TO  FIND A *
3050  REM * VALUE FROM A TABLE *
3060  REM * OF VALUES.         *
3070  REM *                    *
3080  REM *   PARAMETERS:      *
3090  REM *                    *
3100  REM *   X(I)  - ARRAY OF *
3110  REM *           ARGUMENT *
3120  REM *           VALUES OF*
3130  REM *           THE TABLE*
3140  REM *                    *
3150  REM *   Y(I)  - ARRAY OF *
```

```
3160  REM *             FUNCTION *
3170  REM *             VALUES OF*
3180  REM *             THE TABLE*
3190  REM *                      *
3200  REM *   M       - DIMENSION*
3210  REM *             OF TABLE *
3220  REM *                      *
3230  REM *   XV      - INPUT    *
3240  REM *             ARGUMENT *
3250  REM *                      *
3260  REM *   YV      - OUTPUT   *
3270  REM *             VALUE    *
3280  REM *             FOUND    *
3290  REM *                      *
3300  REM *********************
3310 :
3320 :
3330  FOR I = 0 TO M
3340 YY(I) = Y(I)
3350  NEXT I
3360 :
3370  FOR OD = 1 TO M
3380  FOR I = OD TO M
3390 YY(I) = (YY(OD - 1) * (X(I) -
     XV) - YY(I) * (X(OD - 1) - X
     V)) / (X(I) - X(OD - 1))
3400  NEXT I
3410  PRINT "AFTER ITERATION ";OD
     ;" YV = ";YY(OD)
3420  NEXT OD
3430 YV = YY(M)
3440  PRINT
3450  RETURN
```

The output of this program is as follows:

```
--------------------
 X          Y
--------------------
10       .17365
20       .34202
30       .5
40       .64279
50       .76604
60       .86603
--------------------
AFTER ITERATION 1 YV = .392531
AFTER ITERATION 2 YV = .39050495
AFTER ITERATION 3 YV = .39072335
AFTER ITERATION 4 YV = .390732053
AFTER ITERATION 5 YV = .390730174

THE VALUE AT X=23
IS Y=.390730174
```

To reach this output, an Apple II computer required less than 2 seconds. This program has been designed to print the intermediate interpolated values. The reader will note that the accuracy of these values increases as more iterative terms are used to calculate the interpolated value.

6.5 INVERSE INTERPOLATION

Inverse interpolation is the process whereby one finds the value of x corresponding to a given function value y. In this case, y is a value between two given values in the table of data. In such a situation one could invert the table by interchanging the roles of x and y. The only disadvantage to this procedure is that the arguments of the table are no longer uniformly spaced. For this reason, methods that are based on uniform spacing cannot be used.

6.6 CURVE FITTING BY THE METHOD OF LEAST SQUARES

In fitting tabular data with an approximating function, there are two basic schemes. The first scheme requires that the approximating function (perhaps a piecewise function) pass through every point in the table. The interpolation methods discussed in the previous section satisfy this requirement. The alternate approach is to find a simple function that applies over the total range of the table but does not exactly satisfy every data point. This category of problems is called *curve fitting* and seeks to minimize the error between the simple function and the tabular data values. The usual approach to this problem is to seek an approximating function such that the sum of the squares of the difference between the function and the actual data is a minimum. This procedure is known as the method of least squares.

The Method of Least Squares

If one is given $n + 1$ data points $(x_0, y_0) \cdots (x_n, y_n)$ and desires to use an approximation function $g(x)$ on the range

$$x_0 \leqslant x \leqslant x_n$$

the functional error at any tabular point will be

$$\text{Error}_i = g(x_i) - y_i$$

The sum of squares of these individual errors will be

$$E = \sum_{i=0}^{n} [g(x_i) - y_i]^2$$

It is customary to choose $g(x)$ so that it is a linear combination of suitable terms of the form

$$g(x) = c_1 g_1(x) + c_2 g_2(x) \cdots c_k g_k(x)$$

To have a minimum value for E,

$$\frac{\partial E}{\partial c_1} = \frac{\partial E}{\partial c_2} = \cdots = \frac{\partial E}{\partial c_k} = 0$$

Since

$$E = \sum_{i=0}^{n} [c_1 g_1(x_i) + c_2 g_2(x_i) \cdots c_k g_k(x_i) - y_i]^2$$

the requirement of minimum error will give rise to the equation system

$$\frac{\partial E}{\partial c_1} = 2[c_1 g_1(x_i) \cdots + c_k g_k(x_i) - y_i]\; g_1(x_i) = 0$$

$$\vdots$$

$$\frac{\partial E}{\partial c_k} = 2[c_1 g_1(x_i) \cdots + c_k g_k(x_i) - y_i]\; g_k(x_i) = 0$$

Clearly, these k equations can be put into the form

$$\begin{bmatrix} \sum g_1^2(x_i) & \sum g_1(x_i)\, g_2(x_i) \cdots & \sum g_1(x_i)\, g_k(x_i) \\ \vdots & & \\ \sum g_k(x_i)\, g_1(x_i) & \cdots & \sum g_k^2(x_i) \end{bmatrix} \begin{bmatrix} c_1 \\ \vdots \\ c_k \end{bmatrix} = \begin{bmatrix} \sum g_1(x_i)\, y_i \\ \vdots \\ \sum g_k(x_i)\, y_i \end{bmatrix}$$

Since the coefficients in the matrix on the left and in the vector on the right can be determined from the tabular data, this system of k linear equations and k unknowns can be solved. As long as $g(x)$ is linear in its coefficients, any reasonable function may be used. The actual selection should be based on individual judgment of the data. Special characteristics of the data that influence this judgment are periodicity, exponential or log tendencies, symmetry, and asymptotic behavior.

Although the procedure should be used with care, it is possible to break the tabular data into a few separate regions and then do a curve fit for each range. Such a procedure is often justified if the physical situation suggests that a transition from one domain to another has occurred. Examples of this situation in engineering problems might include the transition between laminar and turbulent flow, the transition between subsonic and supersonic flow, and the transition between prebuckling and postbuckling behavior. The engineer who utilizes curve fitting should be careful not to use the approximation formula outside the range of approximation.

Orthogonal Polynomials

If in formulating the approximation function the $g_i(x)$ functions are orthogonal polynomials such that

$$g_j(x_i)\, g_k(x_i) = 0, \qquad j \neq k$$

the system of linear equations found previously will be simplified to a pure diagonal form. Thus the coefficients will be

$$c_j = \frac{\sum_{i=0}^{n} g_j(x_i)\, y_i}{\sum_{i=0}^{n} g_j^2(x_i)}$$

In view of the power of this simplification, it is not surprising that many computer curve fitting subroutines rely on the use of orthogonal polynomials.

The topic of curve fitting by the method of least squares can be illustrated by means of an example application.

EXAMPLE 6-4 The tendency of some normally ductile materials to behave in a brittle matter in the presence of notches is called *notch sensitivity*. This sensitivity is strongly related to temperature and is generally tested by means of a pendulum impact test known as the *Charpy test*. This test measures the impact energy absorbed by a standard test specimen when subjected to test. The results of a Charpy test on AISI C1020 cold-rolled steel are presented in the following table. Suppose that it is desired to find a least-squares polynomial to describe these data over the range of experimental tests.

Experimental Test Results for AISI C1020 Cold-Rolled Steel

Temperature (°*C*)	*Charpy Impact Energy* (*joules*)
−100.	4.06
−75.	6.78
−50.	9.49
−25.	16.27
0.	40.67
25.	97.62
50.	146.43
75.	151.85
100.	162.70

The actual selection of the order of the polynomial to be used for the least-squares fit will depend on the nature of the experimental data. This order must, of course, be less than the total number of data points in the table. Generally, a higher-order polynomial will give better results than a polynomial of lower order. The price one pays for this better performance is that the resulting process will require the solution to an equation system of higher order. This will give rise to a computer program that requires more storage space and requires longer run times to complete. The tendency to use higher-order polynomials should be used with caution since, as the order of the polynomial approaches the maximum allowable value, the resulting solution may tend to zigzag through the data and may thus deviate significantly from the data trend. For this reason, the user may want to try several different polynomial orders to see which gives the most reasonable result. For this example, a polynomial of order 4 has been selected. A BASIC program that implements the least-squares process for this example now follows. This program utilizes an adaptation of the Cholesky method presented in Chapter 3 to solve the resulting simultaneous equations.

```
1000  REM *********************
1010  REM *THIS PROGRAM FINDS  *
1020  REM *A POLYNOMIAL APPROX-*
1030  REM *IMATION TO A SET OF *
1040  REM *DATA VALUES BY THE  *
1050  REM *METHOD OF LEAST     *
1060  REM *SQUARES             *
1070  REM *********************
1080 :
1090 :
1100  REM **SET UP THE TABLE
1110 :
1120 N = 9:KK = 5
1130  DIM X(10),Y(10)
1140  DIM A(6,7),C(6)
1150  PRINT "THE INPUT DATA VALUE
      S"
1160  PRINT "-------------------
      "
1170  PRINT "    T      ENERGY"
```

```
1180  PRINT "  DEG C    JOULES"
1190  PRINT "--------------------
     "
1200  FOR I = 1 TO N
1210  READ X(I),Y(I)
1220  PRINT X(I); TAB( 10);Y(I)
1230  NEXT I
1240  DATA  -100.,4.06,-75.,6.78
1250  DATA  -50.,9.49,-25.,16.27
1260  DATA  0.,40.67,25.,97.62
1270  DATA  50.,146.43,75.,151.85

1280  DATA  100.,162.70
1290  PRINT "--------------------
     "
1300 :
1310  REM **FIND THE COEFFICIENTS

1320 :
1330  GOSUB 3000
1340 :
1350 :
1360  REM **WRITE THE ANSWER**
1370 :
1380  PRINT : PRINT "FOR A POLYNO
     MIAL"
1390  PRINT "OF ORDER ";KK - 1
1400  PRINT "THE COEFFICIENTS ARE
     "
1410  PRINT "--------------------
     "
1420  FOR I = 1 TO KK
1430  PRINT "C(";I;")=";C(I)
1440  NEXT I
1450  PRINT "--------------------
     "
1460 :
1470  END
1480 :
3000  REM **********************
3010  REM * FINDS A POLYNOMIAL *
3020  REM * APPROXIMATION TO   *
3030  REM * A SET OF DATA      *
3040  REM * POINTS IN THE LEAST*
3050  REM * SQUARES SENSE.     *
3060  REM *                    *
3070  REM * THE RESULTING      *
3080  REM * POLYNOMIAL WILL BE *
3090  REM * OF THE FORM:       *
3100  REM *                    *
3110  REM * Y =  C(1) + C(2)*X *
3120  REM *      +C(3)*X^2     *
3130  REM *      +C(4)*X^3 . . .*
3140  REM *                    *
3150  REM *   PARAMETERS:      *
3160  REM *                    *
3170  REM *   X(I)  - ARRAY OF *
3180  REM *           ARGUMENT *
3190  REM *           VALUES OF*
3200  REM *           THE INPUT*
3210  REM *           DATA     *
3220  REM *           STORED   *
3230  REM *           FROM X(1)*
3240  REM *           TO  X(N) *
3250  REM *                    *
3260  REM *                    *
3270  REM *   Y(I)  - ARRAY OF *
3280  REM *           FUNCTION *
3290  REM *           VALUES OF*
3300  REM *           THE INPUT*
3310  REM *           DATA     *
3320  REM *           STORED   *
3330  REM *           FROM Y(1)*
3340  REM *           TO Y(N)  *
3350  REM *                    *
3360  REM *   N     - NUMBER OF*
3370  REM *           DATA     *
3380  REM *           POINTS   *
3390  REM *                    *
3400  REM *   C(I)  - ARRAY OF *
3410  REM *           COEFFI-  *
3420  REM *           CIENTS TO*
3430  REM *           BE FOUND *
3440  REM *                    *
3450  REM *   KK    - NUMBER OF*
3460  REM *           COEFFI-  *
3470  REM *           CIENTS   *
3480  REM *                    *
3490  REM *                    *
3500  REM **********************
3510 :
3520 :
3530  REM ** LOAD THE A-ARRAY
3540  FOR L = 1 TO KK
3550  FOR M = 1 TO KK
3560 S1 = 0.:S2 = 0.
3570  FOR I = 1 TO N
3580 S1 = S1 + X(I) ^ (L - 1) * X
     (I) ^ (M - 1)
3590 S2 = S2 + X(I) ^ (L - 1) * Y
     (I)
3600  NEXT I
3610 A(L,M) = S1
3620 A(L,KK + 1) = S2
3630  NEXT M
3640  NEXT L
3650 :
3660  REM ** SOLVE THE EQUATION
3670  REM ** SYSTEM BY THE
3680  REM ** CHOLESKY METHOD
3690 :
3700 NROW = KK:NCOL = KK + 1
3710  GOSUB 4000
3720 :
3730  RETURN
4000  REM **********************
4010  REM * THIS SUBROUTINE    *
4020  REM * APPLIES CHOLESKY'S *
4030  REM * METHOD TO A        *
4040  REM * MATRIX OF THE FORM *
```

```
4050  REM *                    *
4060  REM *   A(NROW,NCOL)     *
4070  REM *                    *
4080  REM * USING PARTIAL      *
4090  REM * PIVOTING.          *
4100  REM **********************
4110 :
4120 :
4130  REM **USE LARGEST   **
4140  REM **PIVOT ELEMENT **
4150 :
4160  FOR K = 1 TO NROW
4170 :
4180 :
4190  REM **FIND LARGEST PIVOT**
4200 :
4210 PIVOT = A(K,K):IL = K
4220  FOR L = K + 1 TO NROW
4230  IF  ABS (A(L,K)) <  ABS (PI
     VOT) THEN  GOTO 4260
4240 PIVOT = A(L,K)
4250 IL = L
4260  NEXT L
4270  IF IL = K THEN  GOTO 4380
4280 :
4290 :
4300  REM **TRADE ROWS TO GET**
4310  REM **LARGEST PIVOT    **
4320 :
4330  FOR LL = 1 TO NCOL
4340 TEMP = A(K,LL)
4350 A(K,LL) = A(IL,LL)
4360 A(IL,LL) = TEMP
4370  NEXT LL
4380  NEXT K
4390 :
4400 :
4410  REM **CALCULATE FIRST ROW
4420 :
4430  FOR J = 2 TO NCOL
4440 A(1,J) = A(1,J) / A(1,1)
4450  NEXT J
4460 :
4470 :
4480  REM **DO THE ROWS & COLS
4490 :
4500  FOR L = 2 TO NROW
4510 :
4520 :
4530  REM **DO THE LTH COLUMN
4540 :
4550  FOR I = L TO NROW
4560 SUM = 0.
4570  FOR K = 1 TO L - 1
4580 SUM = SUM + A(I,K) * A(K,L)
4590  NEXT K
4600 A(I,L) = A(I,L) - SUM
4610  NEXT I
4620 :
4630 :
4640  REM **DO THE LTH ROW**
4650 :
4660  FOR J = L + 1 TO NCOL
4670 SUM = 0.
4680  FOR K = 1 TO L - 1
4690 SUM = SUM + A(L,K) * A(K,J)
4700  NEXT K
4710 A(L,J) = (A(L,J) - SUM) / A(
     L,L)
4720  NEXT J
4730  NEXT L
4740 :
4750 :
4760  REM **GET C(I) VALUES BY**
4770  REM **BACK SUBSTITUTION **
4780 :
4790 C(NROW) = A(NROW,NCOL)
4800  FOR M = 1 TO NROW - 1
4810 I = NROW - M
4820 SUM = 0.
4830  FOR J = I + 1 TO NROW
4840 SUM = SUM + A(I,J) * C(J)
4850  NEXT J
4860 C(I) = A(I,NCOL). - SUM
4870  NEXT M
4880  RETURN
```

The output of this program is as follows:

```
THE INPUT DATA VALUES
--------------------
    T      ENERGY
  DEG C    JOULES
--------------------
-100      4.06
-75       6.78
-50       9.49
-25       16.27
0         40.67
25        97.62
50        146.43
75        151.85
100       162.7
--------------------
```

```
FOR A POLYNOMIAL
OF ORDER 4
THE COEFFICIENTS ARE
--------------------
C(1)=49.2061305
C(2)=1.46923973
C(3)=.0103986947
C(4)=-7.04053876E-05
C(5)=-7.12085786E-07
--------------------
```

To reach this output, an Apple II computer required approximately 30 seconds. To illustrate the least-squares polynomial found by this process, the accompanying figure illustrates both the experimental data and the curve found from the coefficients in the preceding computer output.

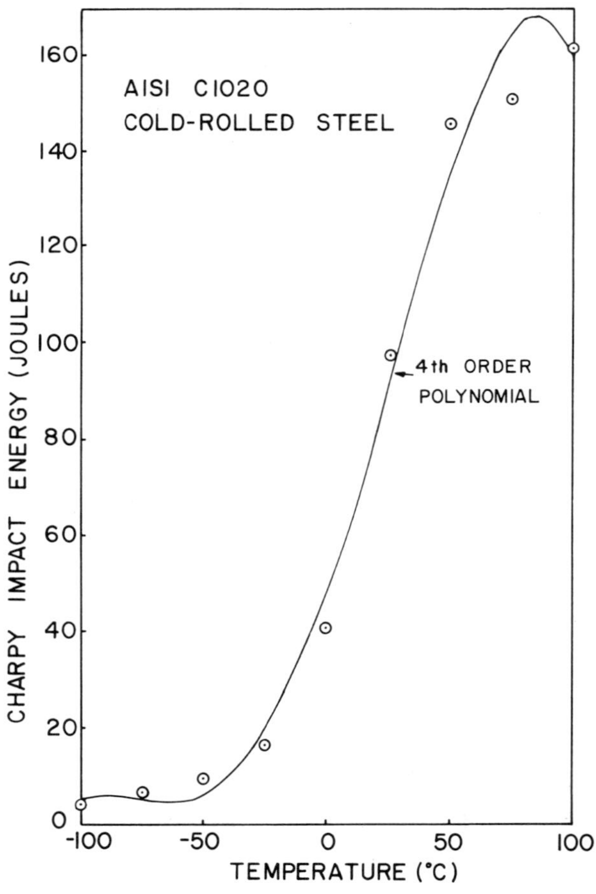

6.7 CURVE SMOOTHING WITH SPLINE FUNCTIONS

Although spline functions are a recent mathematical tool, the physical basis for their development is well established in engineering drafting. A *spline* is a flexible strip or ruler that is constrained to pass through a given set of points (x_i, y_i). When it is constrained in this way, the spline assumes a shape that minimizes its stored strain energy. Based on the assumption of small deflections, the theory of beam deflections can be used to demonstrate mathematically that the spline is a connected group of cubic polynomials arranged so that adjacent curves join each other with continuous first and second derivatives. Such functions are called *cubic* spline functions. To construct a cubic spline, it is necessary to specify the coefficients that uniquely describe each cubic polynomial between given data points. For example, in Figure 6-2 it is necessary to define each of the cubic functions $q_1(x), q_2(x), \ldots, q_m(x)$. In their most general form, these polynomials will be

$$q_i(x) = k_{1i} + k_{2i}x + k_{3i}x^2 + k_{4i}x^3, \qquad i = 1, 2, \ldots, m$$

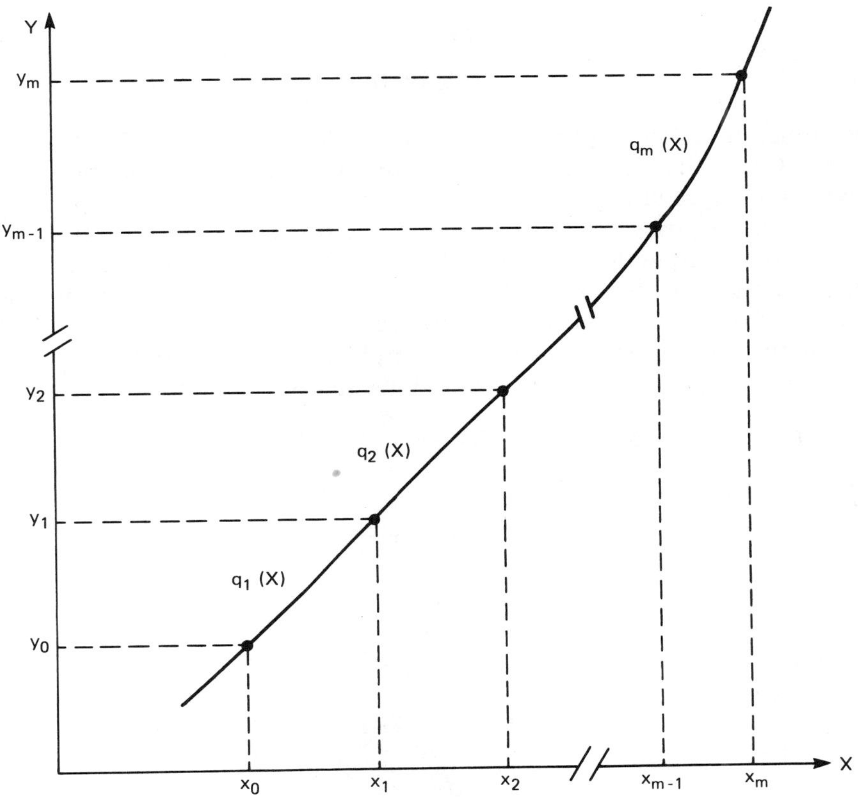

Figure 6-2 Spline functions.

where the k_{ij} terms are constants to be determined by applying the constraint conditions mentioned previously.

The first $2m$ constraint conditions are that the splines must join each other at the data points. Thus

$$q_i(x_i) = y_i, \qquad i = 1, \ldots, m$$

$$q_{i+1}(x_i) = y_i, \qquad i = 0, \ldots, m-1$$

The next $2m - 2$ conditions are that the first and second derivatives must match at all internal data points. Thus

$$q'_{i+1}(x_i) = q'_i(x_i), \qquad i = 1, \ldots, m-1$$

$$q''_{i+1}(x_i) = q''_i(x_i), \qquad i = 1, \ldots, m-1$$

To solve any system of algebraic equations, it is necessary that the number of equations be exactly equal to the number of unknowns. At this point there are $4m$ unknowns and $4m - 2$ constraint equations. This indicates that two more constraints are required. The usual choice for the two additional constraints is that

$$q''_1(x_0) = 0$$

and

$$q''_m(x_m) = 0$$

This choice leads to a spline representation known as the *natural cubic spline.* Once the coefficients associated with a spline function are known, the piecewise polynomial function may be used to represent the data for purposes of interpolation, curve fitting, or surface fitting.

At first inspection the problem of determining the necessary coefficients might seem to be that of solving $4m$ equations for $4m$ unknowns. Fortunately, however, with careful selection of the general form of the cubic polynomials, the complexity of the problem can be greatly simplified. If the individual cubic equations are selected to be

$$q_i(x) = ty_i + \bar{t}y_{i-1} + \Delta x_i[(k_{i-1} - d_i)\,t\bar{t}^2 - (k_i - d_i)\,t^2\bar{t}], \qquad i = 1, \ldots, m$$

where

$$\Delta x_i = x_i - x_{i-1}$$

$$t = \frac{x - x_{i-1}}{\Delta x_i}$$

$$\bar{t} = 1 - t$$

$$\Delta y_i = y_i - y_{i-1}$$

$$\frac{\Delta y_i}{\Delta x_i} = d_i$$

then each of the $q_i(x)$ equations has only two constant coefficients to be determined. After the first $q_i(x)$ equation has been written, each new equation adds only one new constant coefficient to be determined. For this formulation, the value of $t = 0$ and $t = 1$ for $x = x_{i-1}$, and $t = 0$ and $t = 1$ for $x = x_i$. Thus all conditions except for second derivatives are automatically satisfied by this special formulation. The second derivative requirements give rise to the relationships

$$k_{i-1}\,\Delta x_{i+1} + 2k_i(\Delta x_i + \Delta x_{i+1}) + k_{i+1}\,\Delta x_i = 3\,(d_i\,\Delta x_{i+1} + d_{i+1}\,\Delta x_i)$$

for internal points and

$$2k_0 + k_1 = 3d_1$$

and

$$k_{m-1} + 2k_m = 3d_m$$

for the two external points.

Thus the system of equations to be solved is a linear, tridiagonal system of the form

$$\begin{bmatrix} 2 & 1 & 0 & 0 & \\ \Delta x_2 & 2(\Delta x_1 + \Delta x_2) & \Delta x_1 & & \\ & \Delta x_3 & 2(\Delta x_2 + \Delta x_3) & x_2 & \\ & & & & \\ & & \Delta x_m & 2(\Delta x_{m-1} + \Delta x_m) & \Delta x_{m-1} \\ & & & 1 & 2 \end{bmatrix}$$

$$\cdot \begin{bmatrix} k_0 \\ k_1 \\ k_2 \\ \vdots \\ k_m \end{bmatrix} = 3 \begin{bmatrix} d_1 \\ d_1\,\Delta x_2 + d_2\,\Delta x_1 \\ d_2\,\Delta x_3 + d_3\,\Delta x_2 \\ \vdots \\ d_{m-1}\,\Delta x_m + d_m\,\Delta x_{m-1} \\ d_m \end{bmatrix}$$

Because this system is diagonally dominant, an elimination method can be used to find the solution without the need for pivoting. For this equation system the number of coefficients to be determined is equal to the number of data points. Thus the solution process is no more involved than that of fitting the $m + 1$ data points with an mth-order polynomial. Yet it is not uncommon to find that a cubic spline function will do a better job of approximating the function than a polynomial of order m. It is worth mentioning that considerable variety exists in the use of spline functions, since it is possible to choose other types of end constraints and since polynomials of order higher than three could be used.

The use of spline smoothing will now be illustrated by means of an example application using the data from Example 6-4.

EXAMPLE 6-5 Suppose that it is desired to fit a natural cubic spline to the data values used for Example 6-4.

Experimental Test Results for AISI C1020 Cold-Rolled Steel

Temperature (°*C*)	*Charpy Impact Energy* (*joules*)
-100.	4.06
-75.	6.78
-50.	9.49
-25.	16.27
0.	40.67
25.	97.62
50.	146.43
75.	151.85
100.	162.70

A BASIC program that finds the coefficients of this cubic spline now follows. This program utilizes an adaptation of the Cholesky method presented in Chapter 3 to solve the resulting simultaneous equations. The adaptation does not contain the statements that perform pivoting. This modification helps the program to run faster and reduces the number of program lines required.

```
1000  REM *********************
1010  REM *THIS PROGRAM FINDS  *
1020  REM *A NATURAL CUBIC     *
1030  REM *SPLINE PASSING      *
1040  REM *THROUGH M+1 POINTS  *
1050  REM *********************
1060 :
1070  REM **SET UP THE TABLE
1080 :
1090 M = 8
1100  DIM X(10),Y(10)
1110  DIM A(9,10),K(9)
1120  PRINT "THE INPUT DATA VALUE
   S"
1130  PRINT "-------------------
   "
1140  PRINT "    T      ENERGY"
1150  PRINT "  DEG C    JOULES"
1160  PRINT "-------------------
   "
1170  FOR I = 0 TO M
1180  READ X(I),Y(I)
1190  PRINT X(I); TAB( 10);Y(I)
1200  NEXT I
1210  DATA  -100.,4.06,-75.,6.78
1220  DATA  -50.,9.49,-25.,16.27
1230  DATA  0.,40.67,25.,97.62
1240  DATA  50.,146.43,75.,151.85
1250  DATA  100.,162.70
1260  PRINT "-------------------
   "
```

```
1270 :
1280  REM **FIND THE COEFFICIENTS

1290 :
1300  GOSUB 3000
1310 :
1320 :
1330 :
1340  REM **WRITE THE ANSWERS**
1350 :
1360  PRINT "------------------"
1370  PRINT "THE SPLINE"
1375  PRINT "COEFFICIENTS ARE"
1380  PRINT "------------------"
1390  FOR I = 0 TO M
1400  PRINT "K(";I;")=";K(I)
1410  NEXT I
1420  PRINT "------------------"
1430  PRINT
1440 :
1450  END
1460 :
1470 :
1480 :
3000  REM *********************
3010  REM * THIS SUBROUTINE    *
3020  REM * FINDS A  CUBIC     *
3030  REM * SPLINE THAT PASSES *
3040  REM * THROUGH A GROUP OF *
3050  REM * DATA VALUES.       *
3060  REM *                    *
3070  REM * THE METHOD USED IS *
3080  REM * CHOLESKY'S METHOD  *
3090  REM * WITHOUT PIVOTING.  *
3100  REM *                    *
3110  REM *   PARAMETERS:      *
3120  REM *                    *
3130  REM *   X(I)  - ARRAY OF *
3140  REM *           ARGUMENT *
3150  REM *           VALUES OF*
3160  REM *           THE INPUT*
3170  REM *           DATA     *
3180  REM *           STORED   *
3190  REM *           FROM X(0)*
3200  REM *           TO X(M). *
3210  REM *                    *
3220  REM *                    *
3230  REM *   Y(I)  - ARRAY OF *
3240  REM *           FUNCTION *
3250  REM *           VALUES OF*
3260  REM *           THE INPUT*
3270  REM *           DATA     *
3280  REM *           STORED   *
3290  REM *           FROM Y(0)*
3300  REM *           TO Y(M). *
3310  REM *                    *
3320  REM *   M+1   - NUMBER OF*
3330  REM *           DATA     *
3340  REM *           POINTS   *
3350  REM *           AND THE  *
3360  REM *           NUMBER OF*
3370  REM *           COEFFI-  *
3380  REM *           CIENTS TO*
3390  REM *           BE FOUND.*
3400  REM *                    *
3410  REM *   K(I)  - ARRAY OF *
3420  REM *           COEFFI-  *
3430  REM *           CIENTS   *
3440  REM *           FOUND.   *
3450  REM *                    *
3460  REM *                    *
3470  REM *********************
3480 :
3490 :
3500 :
3510 :
3520  FOR I = 0 TO M
3530  FOR J = 0 TO M
3540 A(I,J) = 0.
3550  NEXT J
3560  NEXT I
3570 A(0,0) = 2.:A(0,1) = 1.
3580 A(0,M + 1) = 3. * (Y(1) - Y(
     0)) / (X(1) - X(0))
3590 A(M,M - 1) = 1.:A(M,M) = 2.
3600 A(M,M + 1) = 3. * (Y(M) - Y(
     M - 1)) / (X(M) - X(M - 1))
3610 :
3620  FOR I = 1 TO M - 1
3630 A(I,I - 1) = X(I + 1) - X(I)

3640 A(I,I + 1) = X(I) - X(I - 1)

3650 A(I,I) = 2. * (A(I,I - 1) +
     A(I,I + 1))
3660 D1 = (Y(I) - Y(I - 1)) / A(I
     ,I + 1)
3670 D2 = (Y(I + 1) - Y(I)) / A(I
     ,I - 1)
3680 A(I,M + 1) = 3. * (D1 * A(I,
     I - 1) + D2 * A(I,I + 1))
3690  NEXT I
3700 NROW = M:NCOL = M + 1
3710 :
3720 :
3730 :
3740  REM **CALCULATE FIRST ROW
3750 :
3760  FOR J = 1 TO NCOL
3770 A(0,J) = A(0,J) / A(0,0)
3780  NEXT J
3790 :
3800 :
3810  REM **DO THE ROWS & COLS
3820 :
3830  FOR L = 1 TO NROW
3840 :
3850 :
```

```
3860  REM **DO THE LTH COLUMN
3870 :
3880  FOR I = L TO NROW
3890 SUM = 0.
3900  FOR KK = 0 TO L - 1
3910 SUM = SUM + A(I,KK) * A(KK,L
     )
3920  NEXT KK
3930 A(I,L) = A(I,L) - SUM
3940  NEXT I
3950 :
3960 :
3970  REM **DO THE LTH ROW**
3980 :
3990  FOR J = L + 1 TO NCOL
4000 SUM = 0.
4010  FOR KK = 0 TO L - 1
4020 SUM = SUM + A(L,KK) * A(KK,J
     )
4030  NEXT KK
4040 A(L,J) = (A(L,J) - SUM) / A(
     L,L)
4050  NEXT J
4060  NEXT L
4070 :
4080 :
4090  REM **GET K(I) VALUES BY**
4100  REM **BACK SUBSTITUTION **
4110 :
4120 K(NROW) = A(NROW,NCOL)
4130  FOR MM = 1 TO NROW
4140 I = NROW - MM
4150 SUM = 0.
4160  FOR J = I + 1 TO NROW
4170 SUM = SUM + A(I,J) * K(J)
4180  NEXT J
4190 K(I) = A(I,NCOL) - SUM
4200  NEXT MM
4210  RETURN
```

The output of this program is as follows:

```
THE INPUT DATA VALUES
--------------------
    T      ENERGY
  DEG C    JOULES
--------------------
-100      4.06
-75       6.78
-50       9.49
-25       16.27
0         40.67
25        97.62
50        146.43
75        151.85
100       162.7
--------------------
------------------
THE SPLINE
COEFFICIENTS ARE
------------------
K(0)=.113538734
K(1)=.0993225333
K(2)=.140771134
K(3)=.476392931
K(4)=1.69525714
K(5)=2.5045785
K(6)=.977628866
K(7)=.0925060383
K(8)=.604746978
------------------
```

To reach this output, an Apple II computer required approximately 9 seconds. The coefficients have been used to prepare the accompanying figure. As expected, the cubic spline goes through each of the data points.

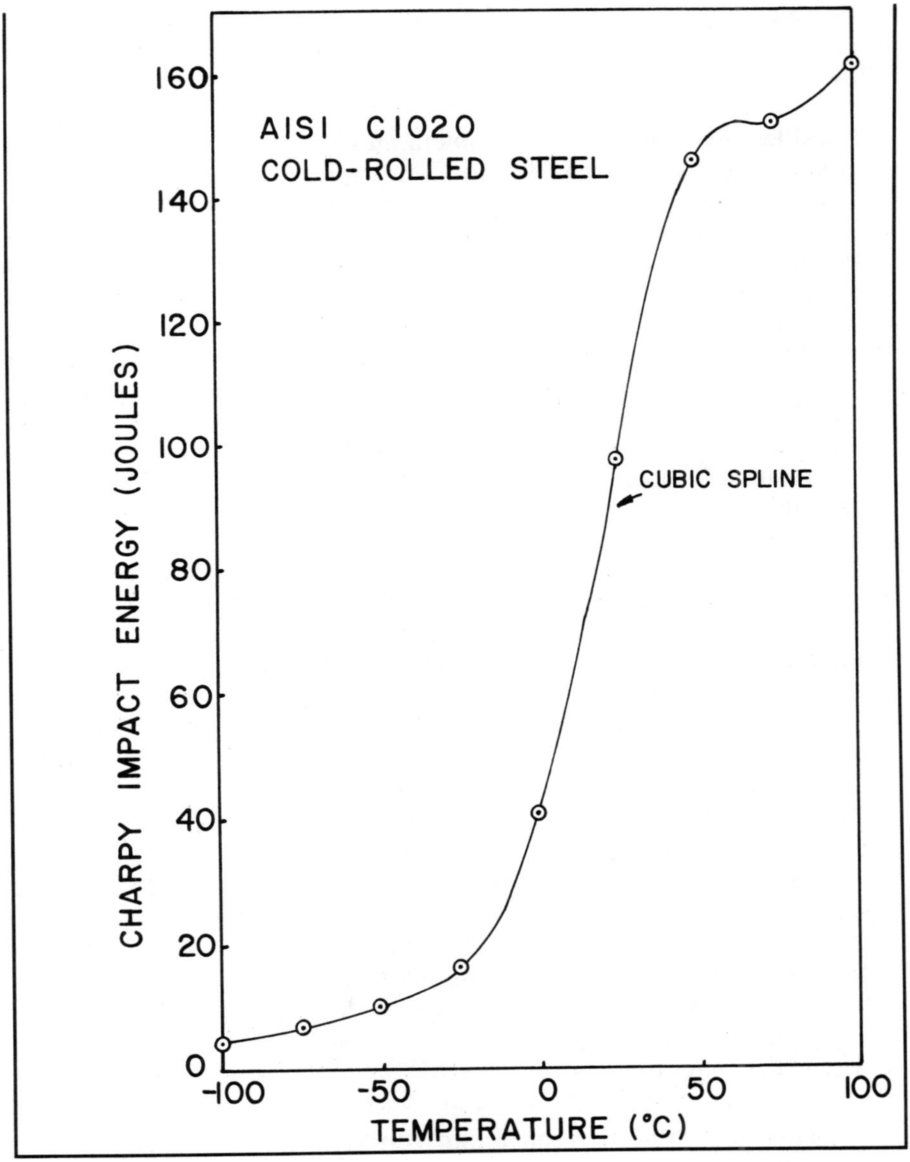

6.8 CONSIDERATIONS IN THE SELECTON OF A METHOD FOR INTERPOLATION, CURVE FITTING, OR SMOOTHING

The selection of an appropriate algorithm for a given interpolation, curve fitting, or smoothing problem will depend strongly on the nature of the data being manipulated and on what type of output is desired.

When selecting an algorithm for use on the small computer, the overall limitations of this device must be kept in mind. These are as follows:

1. **Consider the nature of the problem and its solutions.** Interpolation methods are more plentiful for uniformly spaced data values. If inverse interpolation is to be done or if the data values are nonuniform in spacing, the method of divided differences cannot be used. If the functional values require extensive calculations in order to reach their values, a procedure such as the iterative method can often provide successively better values so that the process can be terminated when the desired degree of accuracy is reached. If multiple values need to be interpolated from a given table of data, a method like that of divided differences can frequently save effort once the initial difference table is assembled.

   Curve fitting by the method of least squares requires that the user select the order of the polynomial to be found. Frequently, information about the nature of the data such as asymptotes, symmetry, or the location of inflections can enable the user to make a better initial selection of the polynomial order. It is often useful to try several different polynomial orders for a given data table and then select the resulting polynomial curve that appears to best describe the desired trend. It should be kept in mind that zigzag behavior of a polynomial curve will often occur as the order of the polynomial approaches the maximum allowable for a given set of data values, since this forces the resulting curve to pass exactly through each of the data points. For this case it is usually better to use a spline curve.

   Some types of data are better represented on logarithmic scales. When this happens, the curve-fitting methods presented in this chapter can still be used by converting the tabular values to their log equivalents prior to applying the least-squares method.

   Some types of data are better represented by a set of piecewise continuous curves rather than by a single curve. This happens when the scientific or engineering phenomenon being measured undergoes a transition from one physical domain to another.
2. **Consider the computer space and run time required.** There is frequently a tradeoff between computer accuracy and computer run time and storage space that will determine which method must be used for interpolation, curve fitting, or data smoothing with the microcomputer. For example, the least-squares process and the curve-smoothing process with splines require the simultaneous solution of linear algebraic equations. As the order of the least-squares approximation goes up, the number of equations in the system also increases. As the number of data points increases, the number of

equations in the spline problem increases. This increase in equation size causes longer computer run times and significantly increases the array storage space needed to implement the solution.

3. **Consider the intermediate outputs of the computer.** Even though the process of outputting information to a CRT screen or to a printer may slow down a numerical solution somewhat, this intermediate information concerning the progress of the technique toward achieving a solution can often give the user considerable information about the nature of the solution process and is thus often justified.
4. **Consider the accuracy required.** Care should be exercised in using interpolated values and in using curves generated from least-squares fits and spline fits since these values and curves can be no more accurate than the input data from which they came. It is also unwise to use information of this type outside the range of the original data.

REFERENCES

1. AHLBERG, H.J., E.N. NILSON, and J.L. WALSH, *The Theory of Splines and Their Application*, Academic Press, Inc., New York, 1967.
2. GROVE, WENDELL E., *Brief Numerical Methods*, Prentice-Hall, Inc., Englewood Cliffs, N.J., 1966.
3. KETTER, ROBERT L., and SHERWOOD P. PRAWEL, Jr., *Modern Methods of Engineering Computation*, McGraw-Hill Book Co., New York, 1969.
4. LAFARA, ROBERT L., *Computer Methods for Science and Engineering*, Hayden Book Co., Rochelle Park, N.J., 1973.
5. McCALLA, THOMAS R., *Introduction to Numerical Methods and FORTRAN Programming*, John Wiley & Sons, Inc., New York, 1967.
6. PALL, GABRIEL A., *Introduction to Scientific Computing*, Appleton-Century-Crofts, Meredith Corp., New York, 1971.
7. RALSTON, ANTHONY, *A First Course in Numerical Analysis*, McGraw-Hill Book Co., New York, 1965.
8. WILLIAMS, P.W., *Numerical Computation*, Harper & Row, Publishers, Inc., New York, 1972.

The microcomputer can help to reduce experimental data in the laboratory to provide meaningful output from experimental tests. (*Photo courtesy of Apple Computer, Inc.*)

Numerical differentiation and integration 7

The analysis of engineering and scientific data frequently leads to the need to obtain the slope of a curve or the area under a curve. To accomplish these tasks, the user can utilize graphical methods; however, it is not uncommon for the accuracy of these procedures to limit the usefulness of the result. For this reason, numerical procedures for differentiation and integration are necessary. It is the purpose of this chapter to present the fundamentals of numerical differentiation and integration.

7.1 NUMERICAL DIFFERENTIATION

Under certain circumstances it may be necessary to approximate the derivative of tabulated data. When this situation arises, the numerical formulas for approximating derivatives can be derived from a consideration of a Taylor's series expansion or by differentiation of the interpolation formulas developed in Chapter 6. This process amounts to that of fitting a polynomial curve to the data points and then using the derivative of the polynomial as the derivative of the tabular data. To understand the pitfalls associated with this process, it is necessary to consider two potential sources of difficulty.

The first difficulty arises from the nature of tabular information obtained by experimental means. The phenomenon of noise or experimental error is present to some degree in every experimental measurement. Thus the true signal shown in Figure 7-1a would be measured with noise as that shown in Figure 7-1b. If these data are differentiated, the effect of the error is greatly amplified by the differentiation process. The result is shown in Figure 7-1c. If, on the other hand, the data with noise are integrated as shown in Figure 7-1d, the effect of the error is diminished. Thus, numerical integration tends to be a far more stable process than that of numerical differentiation.

The second difficulty that arises in numerical differentiation is illustrated in Figure 7-2. Even though an interpolation polynomial may do an adequate job of describing the tabular function, its higher-order derivatives may be totally different from that of the tabular function. This may be observed in Figure 7-2 by a comparison of the slopes and radii of curvature for the two curves.

To illustrate how numerical differentiation formulas can be obtained from Lagrangian interpolation formulas, a simple example can be used. Suppose that a second-order approximating polynomial of the form

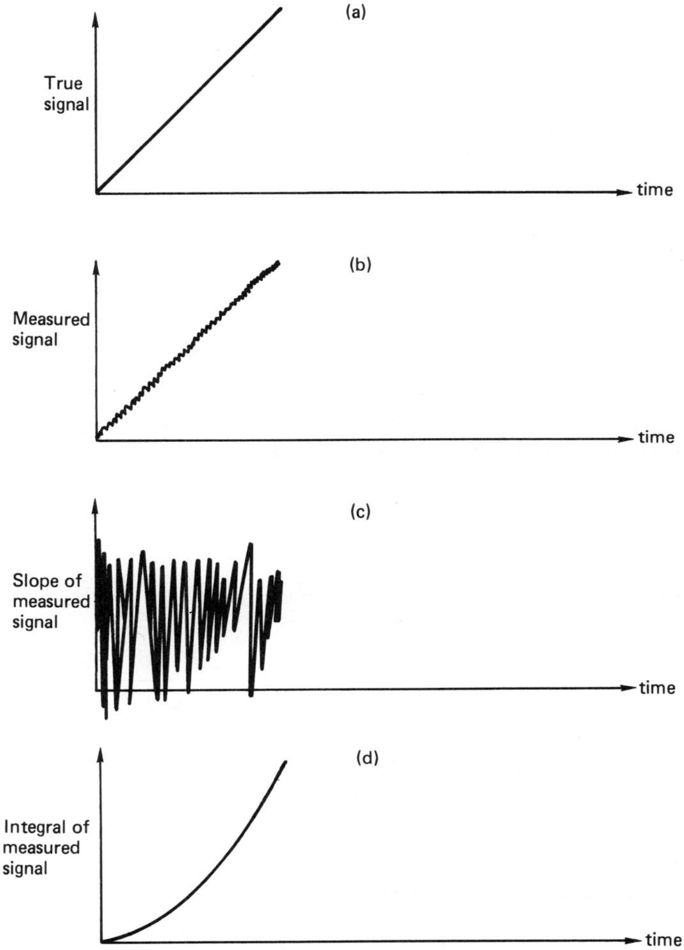

Figure 7-1 Numerical integration and differentiation.

$$P(x) = \frac{(x - x_1)(x - x_2)}{(x_0 - x_1)(x_0 - x_2)} y_0 + \frac{(x - x_0)(x - x_2)}{(x_1 - x_0)(x_1 - x_2)} y_1 + \frac{(x - x_0)(x - x_1)}{(x_2 - x_0)(x_2 - x_1)} y_2$$

is arranged so as to pass through three adjacent points, (x_0, y_0), (x_1, y_1), and (x_2, y_2). The first derivative of this expression is

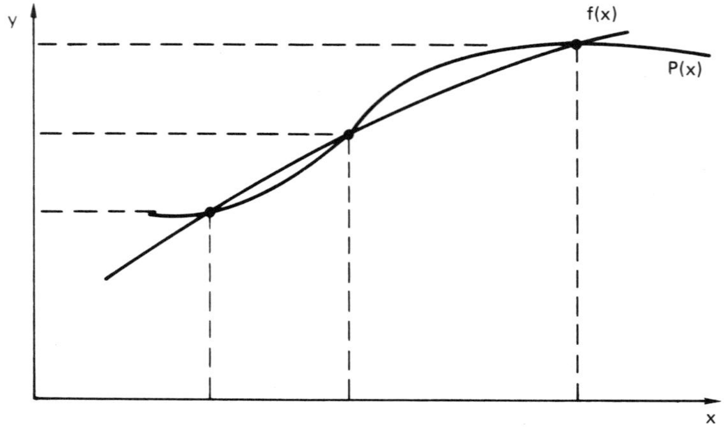

Figure 7-2 A comparison of experimental data and an interpolating polynomial.

$$P'(x) = \frac{2x - x_1 - x_2}{(x_0 - x_1)(x_0 - x_2)} y_0 + \frac{2x - x_0 - x_2}{(x_1 - x_0)(x_1 - x_2)} y_1 + \frac{2x - x_0 - x_1}{(x_2 - x_0)(x_2 - x_1)} y_2$$

If this formula is evaluated at $x = x_0$, the result will be an expression for the first derivative based on forward differences. If the formula is evaluated at $x = x_1$, the result will be an expression for the derivative based on central differences. If the formula is evaluated at $x = x_2$, the result will be an expression for the derivative based on backward differences. Central difference formulas are generally regarded as more accurate than forward or backward difference formulas; however, for some applications central difference formulas cannot be used. An example of this situation occurs at the beginning and end of a table of data.

Other types of approximating polynomials could be used to produce better central, forward, or backward difference approximations to derivatives based on higher-order polynomials using more than three data points. In addition, higher-order derivative approximations could be developed by taking more derivatives of the resulting polynomials.

For many types of engineering and scientific data, the x values are equally spaced. When this occurs, the derivative approximations reduce to expressions that are easier to manipulate. For example, if the spacing $x_1 - x_0 = h$, the forward difference approximation for the derivative at x_0 is

$$P'(x_0) = \frac{(-3y_0 + 4y_1 - y_2)}{2h}$$

Whenever a numerical derivative is used, it is wise to consider the order of magnitude of the error associated with the approximation. To determine this quantity, a Taylor's expansion of the form

$$f(x + \epsilon) = f(x) + \frac{\epsilon}{1} f'(x) + \frac{\epsilon^2}{2!} f''(x) + \frac{\epsilon^3}{3!} f'''(x) + \cdots$$

will be used. If $x = x_0$ and $\epsilon = h$ (one spacing), then $f(x_0 + h) = y_1$. Thus

$$y_1 = y_0 + hD(y_0) + \frac{h^2}{2} D^2(y_0) + \frac{h^3}{6} D^3(y_0) + \cdots$$

In like manner, if $\epsilon = 2h$, the result will be

$$y_2 = y_0 + 2hD(y_0) + 4\frac{h^2}{2} D^2(y_1) + 8\frac{h^3}{6} D^3(y_0) + \cdots$$

If the second-order derivative is eliminated from these two equations by multiplying the first by 4, the second by -1, and then adding, the result will be

$$D(y_0) = \frac{4y_1 - y_2 - 3y_0}{2h} + \frac{h^2 y_0'''}{3}$$

This second-order approximation to the first derivative is the same as found previously except that the error term $\frac{1}{3}h^2 y_0'''$ is present.

A similar procedure to those previously presented could be applied to obtain higher-order derivative approximations for forward, central, or backward forms along with the appropriate error terms. An illustration of the results of this process for different types of interpolation polynomials is presented in Tables 7-1 through 7-3.

Table 7-1 Derivative Approximation Formulas in Terms of Forward Positions

Derivative	*2 Positions*	*3 Positions*	*4 Positions*	*5 Positions*
y_0'	$\frac{1}{h}(y_1 - y_0)$	$\frac{1}{2h}(-y_2 + 4y_1 - 3y_0)$	$\frac{1}{6h}(2y_3 - 9y_2 + 18y_1 - 11y_0)$	$\frac{1}{12h}(-3y_4 + 16y_3 - 36y_2 + 48y_1 - 25y_0)$
	$-\left(\frac{h}{2}y_0''\right)$	$+\left(\frac{h^2}{3}y_0'''\right)$	$-\left(\frac{h^3}{4}y_0^{iv}\right)$	$+\left(\frac{h^4}{5}y_0^{v}\right)$
y_0''		$\frac{1}{h^2}(y_2 - 2y_1 + y_0)$	$\frac{1}{h^2}(-y_3 + 4y_2 - 5y_1 + 2y_0)$	$\frac{1}{12h^2}(11y_4 - 56y_3 + 114y_2 - 104y_1 + 35y_0)$
		$-(hy_0''')$	$+\left(\frac{11h^2}{12}y_0^{iv}\right)$	$+\left(\frac{5h^3}{6}y_0^{v}\right)$
y_0'''			$\frac{1}{h^3}(y_3 - 3y_2 + 3y_1 - y_0)$	$\frac{1}{2h^3}(-3y_4 + 14y_3 - 24y_2 + 18y_1 - 5y_0)$
			$-\left(\frac{3h}{2}y_0^{iv}\right)$	$+\left(\frac{21h^2}{12}y_0^{v}\right)$

Note: The error term is shown in parentheses following each formula.

Table 7-2 Derivative Approximations in Terms of Central Positions

Derivative	*3 Positions*	*5 Positions*	*7 Positions*
y'_0	$\frac{1}{2h}(y_1 - y_{-1})$	$\frac{1}{12h}(-y_2 + 8y_1 - 8y_{-1} + y_{-2})$	$\frac{1}{60h}(y_3 - 9y_2 + 45y_1 - 45y_{-1} + 9y_{-2} - y_{-3})$
	$-\left(\frac{h^2}{6} y'''\right)$	$+\left(\frac{h^4}{30} y_0^{v}\right)$	$-\left(\frac{h^6}{140} y_0^{vii}\right)$
y''_0	$\frac{1}{h^2}(y_1 - 2y_0 + y_{-1})$	$\frac{1}{12h^2}(-y_2 + 16y_1 - 30y_0 + 16y_{-1} - y_{-2})$	$\frac{1}{180h^2}(2y_3 - 27y_2 + 270y_1 - 490y_0 + 270y_{-1} - 27y_{-2} + 2y_{-3})$
	$-\left(\frac{h^2}{12} y^{iv}\right)$	$-\left(\frac{h^4}{90} y_0^{vi}\right)$	$-\left(\frac{h^6}{560} y_0^{viii}\right)$
y'''_0		$\frac{1}{2h^3}(y_2 - 2y_1 + 2y_{-1} - y_{-2})$	$\frac{1}{8h^3}(-y_3 + 8y_2 - 13y_1 + 13y_{-1} - 8y_{-2} + y_{-3})$
		$-\left(\frac{h^2}{4} y_0^{v}\right)$	$+\left(\frac{7h^4}{120} y_0^{vii}\right)$

Note: The error term is shown in parentheses following each formula.

Table 7-3 Derivative Approximations in Terms of Backward Positions

Derivative	*2 Positions*	*3 Positions*	*4 Positions*	*5 Positions*
y_0'	$\frac{1}{h}(y_0 - y_{-1})$	$\frac{1}{2h}(3y_0 - 4y_{-1} + y_{-2})$	$\frac{1}{6h}(11y_0 - 18y_{-1} + 9y_{-2} - 2y_{-3})$	$\frac{1}{12h}(25y_0 - 48y_{-1} + 36y_{-2} - 16y_{-3} + 3y_{-4})$
	$+\left(\frac{h}{2}y_0''\right)$	$+\left(\frac{h^2}{3}y_0'''\right)$	$+\left(\frac{h^3}{4}y_0^{iv}\right)$	$+\left(\frac{h^4}{5}y^{v}\right)$
y_0''		$\frac{1}{h^2}(y_0 - 2y_{-1} + y_{-2})$	$\frac{1}{h^2}(2y_0 - 5y_{-1} + 4y_{-2} - y_{-3})$	$\frac{1}{h^2}(35y_0 - 104y_{-1} + 114y_{-2} - 56y_{-3} + 11y_{-4})$
		$+(hy_0''')$	$+\left(\frac{11h^2}{12}y_0^{iv}\right)$	$+\left(\frac{5h^3}{6}y_0^{v}\right)$
y_0'''			$\frac{1}{h^3}(y_0 - 3y_{-1} + 3y_{-2} - y_{-3})$	$\frac{1}{2h^3}(5y - 18y_{-1} + 24y_{-2} - 14y_{-3} + 3y_{-4})$
			$+\left(\frac{3h}{2}y_0^{iv}\right)$	$+\left(\frac{7h^2}{4}y_0^{v}\right)$

Note: The error term is shown in parentheses following each formula.

EXAMPLE 7-1 As an example illustration of the use of a Lagrangian polynomial to provide an approximation for the derivative of a set of tabular data, the derivative of $y = \sin(x)$ will be developed using 45 equally spaced points from 0° to 90°. A BASIC program that implements this task is presented next. This program first generates the necessary tabular data and then applies a special subroutine to perform the approximation to the derivative. Since the data values are uniformly spaced, the *h* form of the difference equations presented in Tables 7-1 through 7-3 are used. The derivative at 0° is approximated using a forward difference formula, the derivative at 90° is approximated using a backward difference formula, and all other derivatives are approximated using a central difference formula.

```
1000  REM ***********************
1010  REM *THIS PROGRAM FINDS   *
1020  REM *THE DERIVATIVES TO A*
1030  REM *SET OF DATA VALUES   *
1040  REM *BY MEANS OF A        *
1050  REM *LAGRANGIAN INTERPO-  *
1060  REM *LATION POLYNOMIAL    *
1070  REM *FOR UNIFORMLY        *
1080  REM *SPACED DATA VALUES.  *
1090  REM ***********************
1100 :
1110 :
1120  REM **SET UP THE TABLE
1130 :
1140  DIM Y(46),D(46)
1150 M = 45:H = 3.1415926 / 90.
1160  FOR I = 0 TO M
1170 R = I * H
1180 Y(I) =  SIN (R)
1190  NEXT I
1200  GOSUB 3000
1210 :
1220  PRINT "--------------------
     ------------------"
1230  PRINT "ANGLE    SIN(ANGL)
        DERIVATIVE"
1240  PRINT "(DEG)      (UNITS)
        (UNITS)"
1250  PRINT "--------------------
     ------------------"
1260  FOR I = 0 TO M
1265 A = 2 * I
1270  PRINT  TAB( 2);A; TAB( 8);Y
     (I); TAB( 25);D(I)
1280  NEXT I
1290  PRINT "--------------------
     ------------------"
1300  END
1310 :
1320 :
3000  REM **********************
3010  REM * THIS SUBROUTINE     *
3020  REM * COMPUTES A VECTOR   *
3030  REM * OF DERIVATIVES FROM*
3040  REM * A TABLE OF DATA     *
3050  REM * VALUES WITH UNIFORM*
3060  REM * SPACING IN "X."     *
3070  REM *                     *
3080  REM * THE METHOD USED IS  *
3090  REM * BASED ON THE USE OF*
3100  REM * A THREE POINT       *
3110  REM * LAGRANGIAN INTERPO-*
3120  REM * LATION POLYNOMIAL.  *
3130  REM *                     *
3140  REM * THE APPROXIMATION   *
3150  REM * IS BASED ON CENTRAL*
3160  REM * DIFFERENCES EXCEPT  *
3170  REM * FOR THE FIRST AND   *
3180  REM * LAST POINTS.        *
3190  REM *                     *
3200  REM *                     *
3210  REM * PARAMETERS          *
3220  REM *                     *
3230  REM *   H      - SPACING  *
3240  REM *            OF THE X *
3250  REM *            VALUES.  *
3260  REM *                     *
3270  REM *                     *
3280  REM *   Y(I)   - ARRAY OF *
3290  REM *            FUNCTION *
3300  REM *            VALUES OF*
3310  REM *            THE INPUT*
3320  REM *            DATA     *
3330  REM *            STORED   *
3340  REM *            FROM Y(0)*
3350  REM *            TO Y(M). *
```

```
3360  REM *                          *
3370  REM *   M+1     - NUMBER OF*
3380  REM *             DATA       *
3390  REM *             POINTS     *
3400  REM *             AND THE    *
3410  REM *             NUMBER OF*
3420  REM *             DERIVA-    *
3430  REM *             TIVES TO   *
3440  REM *             BE FOUND.*
3450  REM *                          *
3460  REM *   D(I)    - ARRAY OF   *
3470  REM *             DERIVA-    *
3480  REM *             TIVES      *
3490  REM *             FOUND.     *
3500  REM *                          *
3510  REM *                          *
3520  REM ************************
3530 :
3540 :
3550 :
3560 :
3570 D(0) = ( - Y(2) + 4 * Y(1) -
     3 * Y(0)) / (2 * H)
3580 D(M) = (3 * Y(M) - 4 * Y(M -
     1) + Y(M - 2)) / (2 * H)
3590  FOR I = 1 TO M - 1
3600 D(I) = (Y(I + 1) - Y(I - 1))
      / (2 * H)
3610  NEXT I
3620  RETURN
```

The reader will recognize lines 3570, 3580, and 3600 as being the forward, backward, and central difference formulas. The output of this program is as follows:

ANGLE (DEG)	SIN(ANGL) (UNITS)	DERIVATIVE (UNITS)
0	0	1.00040598
2	.0348994961	.999187884
4	.0697564725	.99736148
6	.104528462	.994319942
8	.139173099	.99006698
10	.173648175	.984607773
12	.207911687	.977948972
14	.241921892	.970098692
16	.275637351	.961066498
18	.309016989	.950863392
20	.342020138	.939501804
22	.374606587	.926995578
24	.406736637	.91335995
26	.43837114	.898611534
28	.469471555	.882768293
30	.499999992	.865849549
32	.529919256	.847875899
34	.559192895	.828869226
36	.587785244	.808852716
38	.615661466	.787850745
40	.642787601	.765888891
42	.669130597	.742993927
44	.694658361	.719193739
46	.719339791	.694517319
48	.743144816	.668994739
50	.766044434	.642657092
52	.788010744	.615536465
54	.809016985	.587665904
56	.829037563	.559079364
58	.848048087	.52981167
60	.866025395	.499898486
62	.882947584	.469376247
64	.898794038	.438282144

```
66     .91354545          .406654065
68     .927183847         .374530541
70     .939692614         .341950709
72     .95105651          .308954264
74     .96126169          .275581406
76     .970295721         .241872791
78     .978147596         .207869494
80     .984807749         .173612936
82     .990268065         .139144863
84     .994521893         .104507265
86     .997564049         .0697423324
88     .999390826         .0348923121
90     .999999991         1.02986448E-05
-------------------------------------
```

This output required about 6 seconds on an Apple II computer to complete. The selection of $y = \sin(x)$ is fortunate since the exact answer $y' = \cos(x)$ is known and can be compared with the computer result. The comparison reveals that the approximate result is reasonably accurate. The subroutine in this problem could be applied to a variety of problems in science and engineering as long as equally spaced data values are used. When the spacing is nonuniform, the more general forms of the difference equations must be used. This procedure is illustrated in Example 7-2.

EXAMPLE 7-2 An experimental test has been performed on the action of a piano key for a new linkage design. The result has provided displacement versus time data as follows:

Time (msec)	*Displacement (mm)*
0.00	0.00
5.01	0.18
10.09	1.05
13.98	1.73
16.62	2.35
18.01	2.96
22.53	3.76
25.33	4.48
28.03	5.28
30.42	6.12
32.06	7.09
33.62	8.00

These data come from an experimental test for a particular keystroke. It is desired to find an approximation for the velocity of the key for this test. Since the data are nonuniformly spaced, the subroutine developed in the previous example cannot be used and a new subroutine will be needed. A program that performs this task now follows.

```
1000  REM **********************
1010  REM *THIS PROGRAM FINDS   *
1020  REM *THE DERIVATIVES TO A*
1030  REM *SET OF DATA VALUES   *
1040  REM *BY MEANS OF A        *
1050  REM *LAGRANGIAN INTERPO-  *
1060  REM *LATION POLYNOMIAL    *
1070  REM *FOR NONUNIFORMLY     *
1080  REM *SPACED DATA VALUES.  *
1090  REM **********************
1100 :
1110 :
1120  REM **SET UP THE TABLE
1130 :
1140 M = 11
1150  DIM X(12),Y(12),D(12)
1170  PRINT "--------------------
     ---------"
1180  PRINT " TIME  DISPLACEMENT
      VELOCITY"
1190  PRINT "(MSEC)      (MM)
       (M/SEC)"
1200  PRINT "--------------------
     ---------"
1210  FOR I = 0 TO M
1220  READ X(I),Y(I)
1230  NEXT I
1240  DATA   0.,0.,5.01,0.18,10.0
     9          ,1.05
1250  DATA  13.98,1.73,16.62,2.35
     ,18.01,2.96
1260  DATA  22.53,3.76,25.33,4.48
     ,28.03,5.28
1270  DATA  30.42,6.12,32.06,7.09
     ,33.62,8.
1280  GOSUB 3000
1290  FOR I = 0 TO M
1300  PRINT X(I); TAB( 12);Y(I); TAB(
     19);D(I)
1310  NEXT I
1320  PRINT "--------------------
     ---------"
1330 :
1340  END
1350 :
1360 :
3000  REM **********************
3010  REM * THIS SUBROUTINE     *
3020  REM * COMPUTES A VECTOR   *
3030  REM * OF DERIVATIVES FROM*
3040  REM * A TABLE OF DATA     *
3050  REM * VALUES.             *
3060  REM *                     *
3070  REM * THE METHOD USED IS  *
3080  REM * BASED ON THE USE OF*
3090  REM * A THREE POINT       *
3100  REM * LAGRANGIAN INTERPO-*
3110  REM * LATION POLYNOMIAL.  *
3120  REM *                     *
3130  REM * THE APPROXIMATION   *
3140  REM * IS BASED ON CENTRAL*
3150  REM * DIFFERENCES EXCEPT  *
3160  REM * FOR THE FIRST AND   *
3170  REM * LAST POINTS.        *
3180  REM *                     *
3190  REM *                     *
3200  REM * PARAMETERS          *
3210  REM *                     *
3220  REM *   X(I)  - ARRAY OF  *
3230  REM *           ARGUMENT  *
3240  REM *           VALUES OF*
3250  REM *           THE INPUT*
3260  REM *           DATA      *
3270  REM *           STORED    *
3280  REM *           FROM X(0)*
3290  REM *           TO X(M).  *
3300  REM *                     *
3310  REM *                     *
3320  REM *   Y(I)  - ARRAY OF  *
3330  REM *           FUNCTION  *
3340  REM *           VALUES OF*
3350  REM *           THE INPUT*
3360  REM *           DATA      *
3370  REM *           STORED    *
3380  REM *           FROM Y(0)*
3390  REM *           TO Y(M).  *
3400  REM *                     *
3410  REM *   M+1   - NUMBER OF*
3420  REM *           DATA      *
3430  REM *           POINTS    *
3440  REM *           AND THE   *
3450  REM *           NUMBER OF*
3460  REM *           DERIVA-   *
3470  REM *           TIVES TO  *
3480  REM *           BE FOUND.*
3490  REM *                     *
```

```
3500  REM *    D(I)   - ARRAY OF *
3510  REM *             DERIVA-  *
3520  REM *             TIVES    *
3530  REM *             FOUND.   *
3540  REM *                      *
3550  REM *                      *
3560  REM **********************
3570 :
3580 :
3590 :
3600 :
3610  FOR I = 0 TO M
3620 N = I - 1
3630  IF (I = 0) THEN N = 0
3640  IF (I = M) THEN N = M - 2
3650 D1 = (2 * X(I) - X(N + 1) -
     X(N + 2)) / ((X(N) - X(N + 1
     )) * (X(N) - X(N + 2)))
3660 D2 = (2 * X(I) - X(N) - X(N +
     2)) / ((X(N + 1) - X(N)) * (
     X(N + 1) - X(N + 2)))
3670 D3 = (2 * X(I) - X(N) - X(N +
     1)) / ((X(N + 2) - X(N)) * (
     X(N + 2) - X(N + 1)))
3680 D(I) = D1 * Y(N) + D2 * Y(N +
     1) + D3 * Y(N + 2)
3690  NEXT I
3700  RETURN
```

This program utilizes the more general expressions for forward, backward, and central difference derivative approximations. The output of the program is as follows:

TIME (MSEC)	DISPLACEMENT (MM)	VELOCITY (M/SEC)
0	0	-.0312682697
5.01	.18	.103124557
10.09	1.05	.173268824
13.98	1.73	.210574519
16.62	2.35	.368486486
18.01	2.96	.377261389
22.53	3.76	.226483735
25.33	4.48	.277075518
28.03	5.28	.325560338
30.42	6.12	.493796333
32.06	7.09	.587296748
33.62	8	.579369919

This output required about 3 seconds to complete. The program with its subroutine is more complex than the program in the previous example. This program would not be used for problems having uniform spacing of data values.

7.2 NUMERICAL INTEGRATION

Situations requiring numerical integration often occur in engineering and scientific analysis. For example, when no closed-form solution for an integral can be found or when the complexity of the closed form is quite involved, numerical methods of evaluation are often more practical to use. Another situation in which numerical integration is needed happens when it is desired to find the integral of data gathered in an experimental test. Numerical integration, also called numerical quadrature, is a stable process, as illustrated in Figure 7-1. In contrast to numerical differentiation, numerical integration tends to minimize the influence of data errors on the final result. Numerical integration methods are based on approximating the definite integral as a sum of incremental areas. In general, the problem to be solved can be stated as that of finding the value of

$$I = \int_a^b f(x)\, dx$$

Numerical integration methods are classed according to whether the abscissa values are uniformly spaced or not. The Newton–Cotes formulas require equally spaced abscissa values, but the Gaussian formulas do not. Numerical techniques from each of these two classes will be presented in the following sections.

7.3 THE TRAPEZOIDAL METHOD OF INTEGRATION

The simplest of all Newton–Cotes formulas comes from the trapezoidal method of integration. This method utilizes linear approximations to the function to be integrated. Lines are drawn between adjacent tabulated points (x_i, y_i) and (x_{i+1}, y_{i+1}) for $a \leqslant x \leqslant b$. Thus, if $x_0 = a$ and $x_n = b$, the integral will be represented as the sum of areas of the n trapezoids each having a width of h. The integral value written in terms of the tabular ordinates will be

$$I = \int_a^b f(x)\, dx \cong \frac{h}{2}\,[y_0 + 2y_1 + 2y_2 + \cdots + 2y_{n-1} + y_n]$$

This trapezoidal method is illustrated graphically in Figure 7-3. The method gives a reasonable approximation to the integral if the slivers of area between the top lines of the trapezoids and the curve are small. In general, the size of the sum of these areas will depend on the nature of the function being integrated and on the size of the strip width. The magnitude of this error is known as *truncation error*.

Suppose that a single trapezoidal shape from x_i to x_{i+1} is integrated as

$$I = \int_{x_i}^{x_{i+1}} f(x)\, dx = F(x_{i+1}) - F(x_i)$$

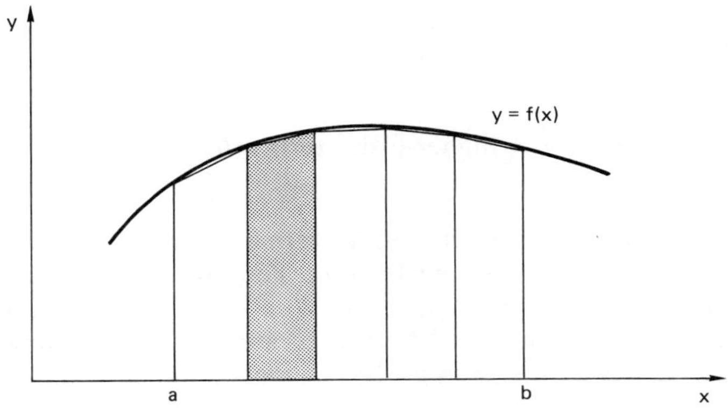

Figure 7-3 Trapezoidal integration.

This integral will be the trapezoidal area plus an error term of the form

$$I = \frac{h[f(x_i + h) + f(x_i)]}{2} + \text{error}$$

The error term can be solved from these two expressions as

$$\text{error} = F(x_{i+1}) - F(x_i) - \frac{h[f(x_i + h) + f(x_i)]}{2}$$

A Taylor's expansion of $F(x_{i+1})$ gives

$$F(x_{i+1}) = F(x_i + h) = F(x_i) + hF'(x_i) + \frac{h^2 F''(x_i)}{2!} + \frac{h^3 F'''(x_i)}{3!} + \text{higher-order terms } (h^4)$$

In this relationship the higher-order terms are on the order of h^4. A Taylor's expansion of $f(x_i + h)$ gives

$$f(x_i + h) = f(x_i) + hf'(x_i) + \frac{h^2 f''(x_i)}{2!} + \frac{h^3 f'''(x_i)}{3!} + \text{higher-order terms } (h^4)$$

In this expression the higher-order terms are on the order of h^4 and are small relative to the other terms. If one recalls that $f(x_i) = F'(x_i)$, $f'(x_i) = F''(x_i)$, and $f''(x_i) = F'''(x_i)$, and substitutes these expressions first into the Taylor's expansion and then uses the result in the error formula, the result will be

$$\text{error} = \frac{-h^3 f''(x_i)}{12} + \text{higher-order terms } (h^4)$$

If the size of the strip width is small, the h^3 term will dominate the higher-order terms and they may be neglected. Thus we see that the inherent truncation error in the trapezoidal rule is dependent on the second derivative of the function being integrated. To determine the error over n intervals, we must sum up these errors from each interval. Thus

$$E = \sum_{i=1}^{n} \text{error} = -\frac{h^3}{12} \sum_{i=1}^{n} f''(x_i)$$

If the second derivative does not change much over the interval being integrated, we can use an average value defined as

$$\bar{f}'' = \frac{1}{n} \sum_{i=1}^{n} f''(x_i)$$

Since $x_{i+n} = x_i + nh$, the total error equation becomes

$$E = \frac{-(x_{i+n} - x_i) h^2 \bar{f}''}{12}$$

Therefore, the total error over an interval (x_i, x_{i+n}) is proportional to the square of the strip width. This means that dividing the interval size in half will reduce the truncation error by a factor of 4. Using this rationale would give the user cause to recommend using a large number of small strips to ensure high accuracy. Unfortunately, there is a limit to the utility of this approach. Another kind of error known as round-off error begins to dominate as the number of strips increases. This type of error arises when many numbers are added to each other. An illustration of the tradeoff between round-off error and truncation error is shown in Figure 7-4. As this figure illustrates, there will be an optimum size for the strip width h. Although the user can experiment with the numerical process to determine a reasonable choice for this optimum value, it is usually impossible to define analytically.

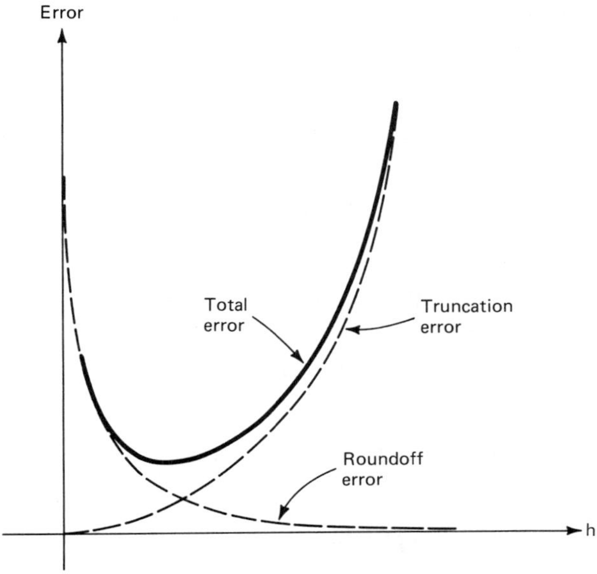

Figure 7-4 The tradeoff between truncation error and round-off error.

7.4 SIMPSON'S METHOD OF INTEGRATION

Another approach to decreasing the error in numerical integration processes is to use some other curve besides a straight line to approximate a small segment of the curve to be integrated. This idea leads quite naturally to the thought that one might use a higher-order curve such as a parabola. Of course, a parabola will require three adjacent tabular points rather than two. As a result, the numerical formula for the integral will differ from that found for trapezoidal segments. This new formula, known as Simpson's rule, will be

$$I = \int_a^b f(x)\,dx \cong \frac{h}{3}\left[y_0 + 4y_1 + 2y_2 + 4y_3 + \cdots + 4y_{n-1} + y_n\right]$$

This equation must be applied to an even number of intervals. By a process similar to that used to find the truncation error for the trapezoidal method, the truncation error for Simpson's method can be found to be

$$E = \frac{-(x_{i+n} - x_i)\bar{f}^{iv} h^4}{180}$$

where $\bar{f}^{iv}$ represents the average value of the fourth derivative of $f(x)$ on the interval (x_i, x_{i+n}). Here we see that the total truncation error is proportional to the fourth power of the interval size. This means that dividing the interval size in half will decrease the truncation error by a factor of 16. This is indeed an improvement over the trapezoidal method.

In using Simpson's rule, it may happen that the number of strips is odd rather than even. This difficulty can be overcome if the first three strips are integrated using the following formula based on a third-degree parabola connecting the first four points on the given curve:

$$A_{3\text{ strips}} = \frac{3h}{8}\,[y_0 + 3y_1 + 3y_2 + y_3]$$

This relationship is known as *Simpson's three-eighths rule.*

7.5 HIGHER-ORDER NEWTON-COTES FORMULAS FOR INTEGRATION

The fact that formulas can be found for polynomials of first and second order leads to the desire to generalize for higher-order polynomials. A detailed mathematical treatment of this situation is presented by Ralston [10]. The result is a general form:

$$\int_a^b f(x)\,dx = c_0 h \sum_{i=0}^{n} w_i f_i + c_1 h^{k+1} f^{(k)}(x^*)$$

where n is the number of strips, k is the order of the polynomial used, x^* is some point in the interval $[a, b]$, $f^{(k)}(x^*)$ is the kth derivative of the function $f(x)$ evaluated at x^*, and h is the width of the data intervals. The coefficients c_0, c_1, and w_i are listed in Table 7-4. Each row of Table 7-4 represents a cycle of k strips involving $k + 1$ pivotal points to achieve a kth-order polynomial. To apply this information to a problem of more than one cycle (say $k = 2$ and $n = 6$), the coefficients should be added end to end so that the last weight values overlap. Thus

	w_0	w_1	w_2	w_3	w_4	w_5	w_6
	1	4	1				
			1	4	1		
					1	4	1
Sum	1	4	2	4	2	4	1

Table 7-4 Coefficients for the Newton-Cotes Formulas

k	c_0	w_0	w_1	w_2	w_3	w_4	w_5	w_6	w_7	w_8	c_1
1	$\frac{1}{2}$	1	1								$\frac{1}{12}$
2	$\frac{1}{3}$	1	4	1							$\frac{1}{90}$
3	$\frac{3}{8}$	1	3	3	1						$\frac{-3}{80}$
4	$\frac{2}{45}$	7	32	12	32	7					$\frac{-8}{945}$
5	$\frac{5}{288}$	19	75	50	50	75	19				$\frac{-275}{12,096}$
6	$\frac{1}{140}$	41	216	27	272	27	216	41			$\frac{-9}{1400}$
7	$\frac{7}{17280}$	751	3577	1323	2989	2989	1323	3577	751		$\frac{-8183}{518,400}$
8	$\frac{4}{14175}$	989	5888	-928	10,946	-4540	10,946	-928	5888	989	$\frac{-2368}{467,775}$

This gives

$$\int_a^b f(x)\,dx = \frac{h}{3}[y_0 + 4y_1 + 2y_2 + 4y_3 + 2y_4 + 4y_5 + y_6] + c_1 h^3 f^{(3)}(x^*)$$

which agrees with the result previously presented. The final term in the Newton–Cotes formula

$$c_1 h^{k+1} f^{(k)}(x^*)$$

is a term that indicates the order of magnitude of the error associated with the approximation. Clearly, as h gets small, h^{k+1} gets even smaller. This trend should be used with care, however, since the error also depends on $f^{(k)}(x^*)$. There do exist some functions whose higher-order derivatives become very large. For these functions the error may not be reduced by increasing the value of k used.

In spite of the attractive appearance of the higher-order Newton–Cotes formulas, they are seldom used because they are unwieldly and have poor round-off characteristics. When very accurate integration techniques are needed, the methods of Romberg integration or Gauss quadrature (to be discussed later in this chapter) are usually employed because they are more efficient and because they are better suited to microcomputer computation.

Before proceeding to these two methods, let us first consider a microcomputer application of Simpson's method

EXAMPLE 7-3 In an industrial assembly-and-machining operation, a special reciprocating device has been developed to provide a concentrated puff of air onto a machined surface to remove machining chips after a machining operation has been completed. It is desired to determine the energy required for the compression stroke portion of the duty cycle of this device. The operation of the device proceeds as shown in the figure. The piston is started at position 1 corresponding to the maximum volume position, and all valves are then closed. As the piston is moved to the minimum volume position shown as position 2, the pressure rises as indicated. At point 2 the outlet valve is opened and the puff of air occurs. During this process, the pressure in the device returns to 1 atmosphere. The piston then continues to move to its starting configuration with the intake valve open. The energy required for the compression stroke of the puff compressor will be the area enclosed on the diagram. This will be the integral of the curve from 1 to 2 minus the area below the line from 3 to 1.

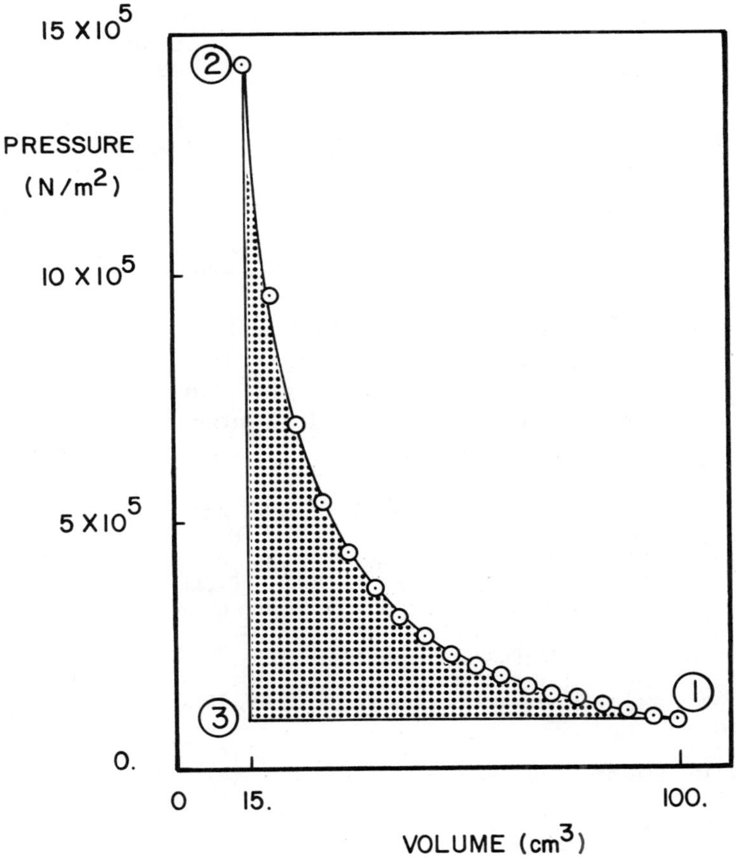

Experimental tests have provided the following data for the compression curve:

Volume (cm^3)	*Pressure* (N/m^2)
15	1,442,700
20	964,400
25	705,600
30	546,700
35	440,600
40	365,400
45	309,900
50	267,400
55	234,000
60	207,100
65	185,200
70	166,900
75	151,600
80	138,500
85	127,200
90	117,400
95	108,900
100	101,300

The method to be used will be Simpson's integration method. The step size h will be 5 cm^3 and there will be an odd number of intervals. Thus the Simpson's three-eighths rule must be used. A program that implements this solution now follows.

```
1000  REM **********************
1010  REM *THIS PROGRAM FINDS   *
1020  REM *THE AREA UNDER A P-V*
1030  REM *CURVE FOR A COMPRES-*
1040  REM *SOR USING SIMPSON'S *
1050  REM *RULE FOR INTEGRATION*
1060  REM *FOR UNIFORMLY        *
1070  REM *SPACED DATA VALUES. *
1080  REM **********************
1090 :
1100 :
1110  REM **SET UP THE TABLE
1120 :
1130  DIM Y(17)
1140 M = 17:H = 5 * 10 ^ ( - 6)
1150 :
1160  FOR I = 0 TO M
1170  READ Y(I)
1180  NEXT I
1190 :
1200  DATA  1442700.,964400.,7056
     00.,546700.
1210  DATA  440600.,365400.,30990
     0.,267400.
1220  DATA   234000.,207100.,1852
     0          0.,166900.
1230  DATA  151600.,138500.,12720
     0.,117400.
1240  DATA  108900.,101300.
1250 :
1260  GOSUB 3000
1270 :
1280  PRINT "--------------------
     ---------"
1290  PRINT "V(CM^3)          P(N/M
     ^2)"
1300  PRINT "--------------------
     ---------"
1310  FOR I = 0 TO M
1320 V = 15 + 5 * I
1330  PRINT V,Y(I)
1340  NEXT I
1350  PRINT "--------------------
     ---------"-
1360 :
1370  REM **REMOVE AREA UNDER
1380  REM **LINE 3--1
1390 B = A - 85 * 10 ^ ( - 6) * 1
     01300.
```

```
1400  PRINT "TOTAL WORK=";B;" JOU
    LES"
1410  END
1420 :
1430 :
3000  REM *********************
3010  REM * THIS SUBROUTINE    *
3020  REM * COMPUTES THE       *
3030  REM * INTEGRAL FOR A     *
3040  REM * TABLE OF EQUALLY   *
3050  REM * SPACED FUNCTION    *
3060  REM * VALUES.            *
3070  REM *                    *
3080  REM * THE METHOD USED IS *
3090  REM * SIMPSON'S RULE WITH*
3100  REM * SIMPSON'S THREE-   *
3110  REM * EIGHTHS RULE USED  *
3120  REM * WHEN THE NUMBER OF *
3130  REM * DATA VALUES IS EVEN*
3140  REM *                    *
3150  REM *                    *
3160  REM * PARAMETERS         *
3170  REM *                    *
3180  REM *   H      - SPACING *
3190  REM *            OF THE X *
3200  REM *            VALUES.  *
3210  REM *                    *
3220  REM *                    *
3230  REM *   Y(I)   - ARRAY OF *
3240  REM *            FUNCTION *
3250  REM *            VALUES OF*
3260  REM *            THE INPUT*
3270  REM *            DATA     *
3280  REM *            STORED   *
3290  REM *            FROM Y(0)*
3300  REM *            TO Y(M). *
3310  REM *                    *
3320  REM *   M+1    - NUMBER OF*
3330  REM *            DATA     *
3340  REM *            POINTS   *
3350  REM *                    *
3360  REM *   A      - INTEGRAL *
3370  REM *            VALUE    *
3380  REM *            FOUND.   *
3390  REM *                    *
3400  REM *                    *
3410  REM *********************
3420 :
3430 :
3440 A = 0:IS = 1
3450  IF (2 *  INT (M / 2) = M) THEN
    GOTO 3490
3460 A = (3 * H / 8) * (Y(0) + 3 *
    Y(1) + 3 * Y(2) + Y(3))
3470 IS = 5
3480  IF M = 3 THEN  GOTO 3540
3490 A = A + (H / 3) * (Y(IS - 1)
    + 4 * Y(M - 1) + Y(M))
3500  IF IS > M - 2 THEN  GOTO 35
    40
3510  FOR I = IS TO M - 2 STEP 2
3520 A = A + (H / 3) * (4 * Y(I) +
    2 * Y(I + 1))
3530  NEXT I
3540  RETURN
```

The reader will recognize statement 3460 as being the Simpson's three-eighths rule and statements 3490 and 3520 as being the normal Simpson's rule. The subroutine in this program is a general-purpose routine that is suitable for any number of even or odd intervals greater than two. The output of this program now follows.

```
-----------------------------
V(CM^3)          P(N/M^2)
-----------------------------
15               1442700
20               964400
25               705600
30               546700
35               440600
40               365400
45               309900
50               267400
55               234000
60               207100
65               185200
70               166900
75               151600
```

```
80              138500
85              127200
90              117400
95              108900
100             101300
------------------------------
TOTAL WORK=18.2832083 JOULES
```

This program required less than 2 seconds to complete on an Apple II computer.

7.6 ROMBERG INTEGRATION

Of all the Newton–Cotes formulas, the trapezoidal rule is the easiest to apply. Unfortunately, the trapezoidal method lacks the degree of accuracy generally desired for scientific or engineering problems. The Romberg integration method has gained widespread acceptance because it combines the simplicity of the trapezoidal method with high accuracy. Romberg's method involves the application of Richardson's extrapolation process to obtain improved results found from the trapezoidal method. To illustrate the Romberg integration process, recall the trapezoidal rule:

$$\int_b^a f(x)\,dx = \frac{h}{2}\left[f(a) + f(b) + 2\sum_{j=1}^{m-1} f(x_j)\right]$$

where $h = (b - a)/m$ and $x_j = a + jh$. It is well known that the truncation error will decrease as the value of m increases. One systematic way to increase m is to use the progression

$$m = 2^{k-1}, \qquad \text{for} \quad k = 1, 2, 3, \ldots, n$$

where n is a positive integer. The values of the step size h for this progression must decrease as

$$h_k = \frac{b - a}{2^{k-1}}, \qquad \text{for} \quad k = 1, 2, 3, \ldots, n$$

It is convenient to express the resulting integral approximation as $R_{k,1}$ so that

$$R_{1,1} = \frac{h_1}{2}\left[f(a) + f(b)\right]$$

$$R_{2,1} = \frac{h_2}{2} [f(a) + f(b) + 2f(a + h_2)]$$

$$R_{3,1} = \frac{h_3}{2} [f(a) + f(b) + 2[f(a + h_3) + f(a + 2h_3) + f(a + 3h_3)]]$$

In general it can be shown that this sequence reduces to

$$R_{k,1} = \frac{1}{2}\left[R_{k-1,1} + h_{k-1} \sum_{i=1}^{2^{k-2}} f(a + (i - 0.5)h_{k-1})\right], \qquad \text{for} \quad k = 2, 3, \ldots, n$$

This relationship can be used to speed the calculation process for each $R_{k,1}$. Although as k increases the accuracy of $R_{k,1}$ also improves, the convergence to an accurate value of the integral is very slow. To speed the convergence process, a Richardson extrapolation can be used:

$$R_{i,j} = \frac{2^{2(j-1)}R_{i,j-1} - R_{i-1,j-1}}{2^{2(j-1)} - 1}, \qquad \text{for each} \quad \begin{array}{l} i = 2, 3, 4, \ldots, n \\ j = 1, 2, \ldots, i \end{array}$$

where the values with larger j index are the extrapolated values and actually correspond to successively higher order Newton–Cotes formulas. The resulting approximations are generally written in the form of a triangular table:

$R_{1,1}$				
$R_{2,1}$	$R_{2,2}$			
$R_{3,1}$	$R_{3,2}$	$R_{3,3}$		
$R_{4,1}$	$R_{4,2}$	$R_{4,3}$	$R_{4,4}$	
$\vdots$				
$R_{n,1}$	$R_{n,2}$	$\cdots$		$R_{n,n}$

It can be shown that the diagonal terms of this table tend to the value of the integral more rapidly than the $R_{n,1}$ terms. The usual procedure to implement the Romberg iteration scheme is to proceed with the generation of values for the rows of this table until a value of n is found in which the difference between $R_{n,n}$ and $R_{n-1,n-1}$ is smaller than some predetermined error tolerance. A logic flow diagram for this procedure is illustrated in Figure 7-5. The procedure is best illustrated by an example application.

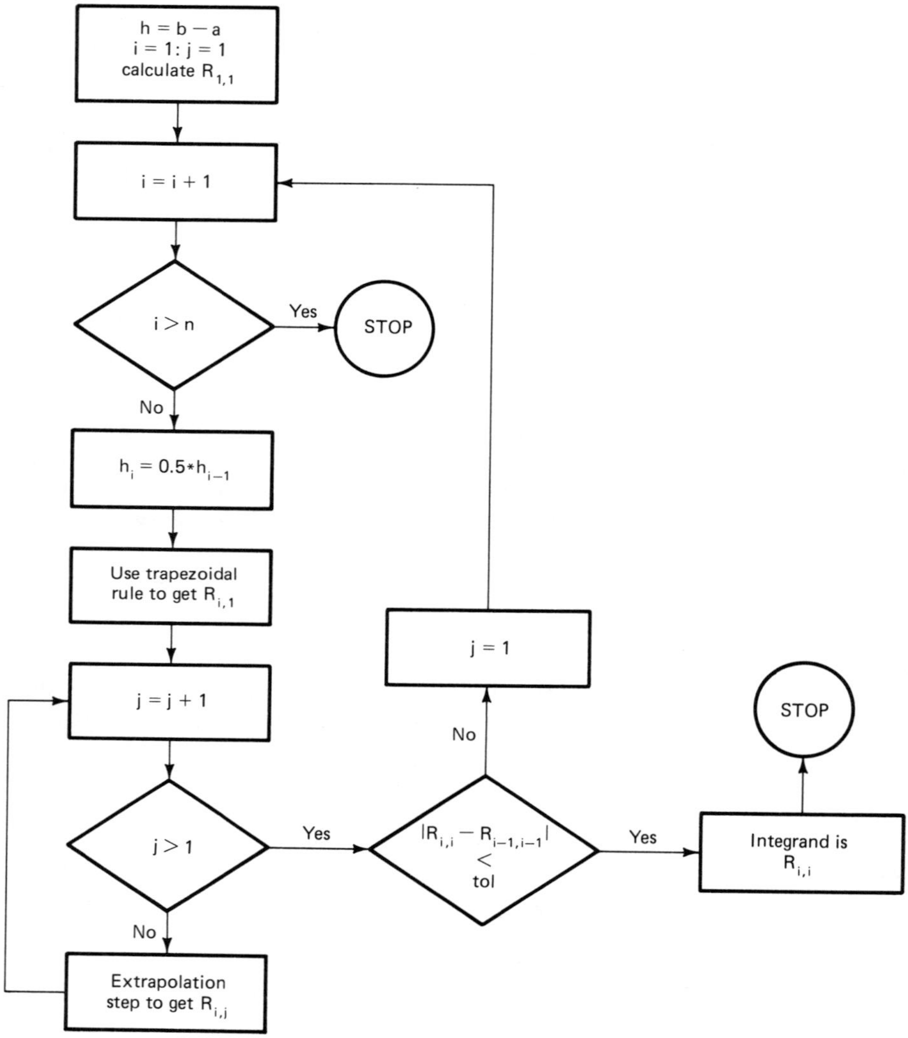

Figure 7-5 Logic flow diagram for the Romberg integration method.

EXAMPLE 7-4 As an illustration of the use of Romberg's method, suppose that it is desired to integrate the function $y = \sin(x)$ from 0 to π. This particular problem was selected because the exact answer is known to be 2.0 and thus the progress of the Romberg method can be monitored. A BASIC program that implements this method is presented next. The program used terminates calculation if the absolute difference between two diagonal elements of Romberg's array is less than 10^{-7} or if more than 7 rows of the array are required.

```
1000  REM **********************
1010  REM *THIS PROGRAM FINDS   *
1020  REM *THE INTEGRAL OF A    *
1030  REM *FUNCTION FN F(X)     *
1040  REM *USING ROMBERG'S      *
1050  REM *INTEGRATION METHOD.  *
1060  REM **********************
1070 :
1080 :
1090  REM ** SET UP THE FUNCTION
1100 :
1110  DEF  FN F(X) =  SIN (X)
1120 :
1130  DIM R(8,8)
1140 A = 0:B = 3.1415926:NMAX = 7

1150 EPS = 10 ^ ( - 7)
1160 :
1170  GOSUB 3000
1180 :
1190  PRINT "INTEGRAL=";Y
1200 :
1210  END
1220 :
1230 :
3000  REM **********************
3010  REM * THIS SUBROUTINE     *
3020  REM * COMPUTES THE        *
3030  REM * INTEGRAL OF AN      *
3040  REM * EXTERNAL FUNCTION   *
3050  REM *                     *
3060  REM * EVALUATION OF THE   *
3070  REM * INTEGRAL IS DONE BY*
3080  REM * THE TRAPEZOIDAL     *
3090  REM * RULE IN CONNECTION *
3100  REM * WITH ROMBERG'S      *
3110  REM * PRINCIPLE.          *
3120  REM *                     *
3130  REM *                     *
3140  REM * PARAMETERS          *
3150  REM *                     *
3160  REM *   A      - LOWER    *
3170  REM *            BOUND OF *
3180  REM *            INTERVAL.*
3190  REM *                     *
3200  REM *   B      - UPPER    *
3210  REM *            BOUND ON *
3220  REM *            INTERVAL.*
3230  REM *                     *
3240  REM *   NMAX   - UPPER    *
3250  REM *            BOUND ON *
3260  REM *            ROMBERG  *
3270  REM *            STEPS TO *
3280  REM *            BE USED. *
3290  REM *                     *
3300  REM * FN F(X) - EXTERNAL  *
3310  REM *            FUNCTION *
3320  REM *            TO BE    *
3330  REM *            INTE-    *
3340  REM *            GRATED.  *
3350  REM *                     *
3360  REM *  EPS    - UPPER     *
3370  REM *            BOUND ON *
3380  REM *            ABSOLUTE *
3390  REM *            ERROR.   *
3400  REM *                     *
3410  REM *  Y      - VALUE     *
3420  REM *            FOUND.   *
3430  REM *                     *
3440  REM *  R(I,J) - WORKING    *
3450  REM *            ARRAY OF *
3460  REM *            ROMBERG  *
3470  REM *            VALUES.  *
3480  REM *                     *
3490  REM **********************
3500 :
3510 :
3520 H = B - A
3530 R(1,1) = 0.5 * H * ( FN F(A)
      -  FN F(B))
3540  FOR I = 2 TO NMAX
3550 SUM = 0.
3560  FOR II = 1 TO 2 ^ (I - 2)
3570 SUM = SUM +  FN F(A + H * (I
     I - .5))
3580  NEXT II
3590 R(I,1) = 0.5 * (R(I - 1,1) +
     H * SUM)
3600 H = 0.5 * H
3610  FOR J = 2 TO I
3620 R(I,J) = (2 ^ (2 * (J - 1)) *
     R(I,J - 1) - R(I - 1,J - 1))
      / (2 ^ (2 * (J - 1)) - 1)
3630  NEXT J
3640  IF  ABS (R(I,I) - R(I - 1,I
      - 1)) < EPS THEN  GOTO 3680

3650  NEXT I
3660  PRINT "NO SOLUTION FOUND"
3670  RETURN
3680 Y = R(I,I)
3690  RETURN
```

The output of this program is the single line

```
INTEGRAL=2
```

This program required less than 4 seconds to complete on an Apple II computer. If the intermediate values of the $R(I, J)$ array are printed, the values will be

```
-8.3875E-08
 1.57079625  2.09439502
 1.89611885  2.00455973  1.99857070
 1.97423158  2.00026916  1.99998312  2.00000554
 1.99357033  2.00001659  1.99999975  2.00000001  1.99999999
 1.99839336  2.00000103  1.99999999  2.00000000  2.00000000  2.00000000
```

The values in the first column of this array represent the answers from trapezoidal integration using progressively smaller step sizes. The reader will note that these values are converging to the correct answer, but the convergence is rather slow when compared to the convergence of the diagonal terms in the array.

7.7 GAUSSIAN QUADRATURE

All Newton–Cotes integration formulas involve the use of equally spaced abscissa points. If this restriction is lifted, it is possible to choose the spacing of the data points in such a way that the error in the approximation is reduced. This is the basis of Gaussian quadrature.

In the formula

$$\int_a^b f(x)\,dx = \sum_{i=0}^{n} w_i f(x_i) + E$$

both w_i and x_i are treated as unknowns to be determined. Thus the total number of unknowns to be determined will be $2(n + 1)$. A polynomial of order $2n + 1$ requires $2n + 2$ conditions to specify its form uniquely. Thus we will approximate the integral using a polynomial of order $2n + 1$ and require that the approximation have zero error for all polynomials of degree less than or equal to this value. This process will give $2n + 2$ equations in $2n + 2$ unknowns. In general, the resulting equations will be linear in w_i but nonlinear in the x_i values. If the limits of integration are $[-1, 1]$, the x_i values will be the $(n + 1)$ roots of the

Legendre polynomial $p_{n+1}(x) = 0$. Once the x_i values are known, the w_i values can be found by linear methods. To illustrate the use of this information, let us consider a problem in which $n = 1$. The requirement that

$$\int_{-1}^{1} f(x)\,dx = \sum_{i=0}^{1} w_i f(x_i)$$

be satisfied for all polynomials of degree $2n + 1$ and less means that if one uses $f(x) = 1, x, x^2$, or x^3 the equality should hold with no error. This gives the following equations:

$$w_0 + w_1 = \int_{-1}^{+1} dx = 2$$

$$w_0 x_0 + w_1 x_1 = \int_{-1}^{+1} x\,dx = 0$$

$$w_0 x_0^2 + w_1 x_1^2 = \int_{-1}^{+1} x^2\,dx = \tfrac{2}{3}$$

$$w_0 x_0^3 + w_1 x_1^3 = \int_{-1}^{+1} x^3\,dx = 0$$

This system of four equations and four unknowns can be solved even though it is nonlinear. To shortcut the nonlinear difficulty, the Legendre polynomial

$$p_2(x) = [-1 + 3x^2]$$

can be used. The advantage to this procedure is that the roots of Legendre polynomials are well known and are tabulated in the scientific literature; see for example the text by Stroud and Secrest [12].

The second-order polynomial will have roots $x_1 = -\sqrt{3}/3$ and $x_2 = +\sqrt{3}/3$. Using these values and any two of the constraint equations shown previously, one can find $w_0 = 1.$ and $w_1 = 1.$ Thus the integral can be expressed as

$$\int_{-1}^{+1} f(x)\,dx = f\left(\frac{-\sqrt{3}}{3}\right) + f\left(\frac{+\sqrt{3}}{3}\right)$$

The restriction that the interval of integration must be $[-1, 1]$ is not as confining as one might think. Clearly, a change of variable

$$z = \frac{2x - (a + b)}{b - a}$$

will convert a general integral to the required form:

$$I = \int_a^b f(x)\,dx = \frac{b - a}{2} \int_{-1}^{+1} f\left(\frac{(b - a)z + b + a}{2}\right) dz$$

Table 7-5 Roots and Coefficients of Legendre Polynomials Used for Gaussian Quadrature

n	Roots	Coefficients
2	0.5773502692	1.0000000000
	−0.5773502692	1.0000000000
3	0.7745966692	0.5555555556
	0.0000000000	0.8888888889
	−0.7745966692	0.5555555556
4	0.8611363116	0.3478548451
	0.3399810436	0.6521451549
	−0.3399810436	0.6521451549
	−0.8611363116	0.3478548451
5	0.9061798459	0.2369268850
	0.5384693101	0.4786286705
	0.0000000000	0.5688888889
	−0.5384693101	0.4786286705
	−0.9061798459	0.2369268850
6	0.9324695142	0.1713244924
	0.6612093864	0.3607615730
	0.2386191861	0.4679139346
	−0.2386191861	0.4679139346
	−0.6612093864	0.3607615730
	−0.9324695142	0.1713244924
7	0.9491079123	0.1294849661
	0.7415311856	0.2797053914
	0.4058451414	0.3818300505
	0.0000000000	0.4179591837
	−0.4058451414	0.3818300505
	−0.7415311856	0.2797053914
	−0.9491079123	0.1294849661
8	0.9602898565	0.1012285363
	0.7966664774	0.2223810345
	0.5255324099	0.3137066459
	0.1834346425	0.3626837834
	−0.1834346425	0.3626837834
	−0.5255324099	0.3137066459
	−0.7966664774	0.2223810345
	−0.9602898565	0.1012285363
9	0.9681602395	0.0812743884
	0.8360311073	0.1806481607
	0.6133714327	0.2606106964
	0.3242534234	0.3123470770
	0.0000000000	0.3302393550
	−0.3242534234	0.3123470770
	−0.6133714327	0.2606106964
	−0.8360311073	0.1806481607
	−0.9681602395	0.0812743884
10	0.9739065285	0.0666713443
	0.8650633667	0.1494513492
	0.6794095683	0.2190863625
	0.4333953941	0.2692667193
	0.1488743390	0.2955242247
	−0.1488743390	0.2955242247
	−0.4333953941	0.2692667193
	−0.6794095683	0.2190863625
	−0.8650633667	0.1494513492
	−0.9739065285	0.0666713443

Thus the Gaussian quadrature approach to the integral of a function using $n = 2$ would give

$$I = \frac{b-a}{2}\left[w_1 f\left(\frac{(b-a)z_1 + b + a}{2}\right) + w_2 f\left(\frac{(b-a)z_2 + b + a}{2}\right)\right]$$

For a function using $n = 3$, the result would be

$$I = \frac{b-a}{2}\left[w_1 f\left(\frac{(b-a)z_1 + b + a}{2}\right) + w_2 f\left(\frac{(b-a)z_2 + b + a}{2}\right) + w_3 f\left(\frac{(b-a)z_3 + b + a}{2}\right)\right]$$

The problem then is to find the coefficients w_i and the roots z_i for the Legendre polynomials. Table 7-5 provides these values for $n = 1$ to 10. This formulation lends itself to computer solution. To illustrate this fact, the problem worked in the previous example problem will be done using Gaussian quadrature.

EXAMPLE 7-5 As an illustration of the use of Gaussian quadrature, suppose that it is desired to integrate the function $y = \sin(x)$ from 0 to π. This problem is the same as was selected for Example 7-4, and the exact answer is known to be 2.0. A BASIC program that implements this method is presented next. The program uses a general-purpose subroutine that utilizes a tenth-order Legendre polynomial.

```
1000  REM **********************
1010  REM *THIS PROGRAM FINDS  *
1020  REM *THE INTEGRAL OF A   *
1030  REM *FUNCTION FN F(X)    *
1040  REM *USING GAUSS QUADRA- *
1050  REM *TURE.               *
1060  REM **********************
1070 :
1080 :
1090  REM ** SET UP THE FUNCTION
1100 :
1110  DEF  FN F(X) =  SIN (X)
1120 :
1130 A = 0:B = 3.1415926
1140 :
1150  GOSUB 3000
1160 :
1170  PRINT "INTEGRAL=";Y
1180 :
1190  END
1200 :
1210 :
3000  REM **********************
3010  REM * THIS SUBROUTINE    *
3020  REM * COMPUTES THE       *
3030  REM * INTEGRAL OF AN     *
3040  REM * EXTERNAL FUNCTION  *
3050  REM *                    *
3060  REM * EVALUATION OF THE  *
3070  REM * INTEGRAL IS DONE BY*
3080  REM * MEANS OF A 10 POINT*
3090  REM * GAUSS QUADRATURE   *
3100  REM * FORMUAL.           *
3110  REM *                    *
3120  REM *                    *
3130  REM * PARAMETERS         *
3140  REM *                    *
3150  REM *   A      - LOWER   *
```

```
3160  REM *              BOUND OF *
3170  REM *              INTERVAL.*
3180  REM *                       *
3190  REM *   B        - UPPER    *
3200  REM *              BOUND ON *
3210  REM *              INTERVAL.*
3220  REM *                       *
3230  REM * FN F(X)    - EXTERNAL *
3240  REM *              FUNCTION *
3250  REM *              TO BE    *
3260  REM *              INTE-    *
3270  REM *              GRATED.  *
3280  REM *                       *
3290  REM *   Y        - VALUE    *
3300  REM *              FOUND.   *
3310  REM *                       *
3320  REM *********************
3330 :
3340 :
3350 D = .5 * (A + B)
3360 E = B - A
3370 C = .4869533 * B
3380 Y = .03333567 * ( FN F(D + C
     ) +  FN F(D - C))
3390 C = .4325317 * B
3400 Y = Y + .07472567 * ( FN F(D
      + C) +  FN F(D - C))
3410 C = .3397048 * B
3420 Y = Y + .1095432 * ( FN F(D +
     C) +  FN F(D - C))
3430 C = .2166977 * B
3440 Y = Y + .1346334 * ( FN F(D +
     C) +  FN F(D - C))
3450 C = .07443717 * B
3460 Y = B * (Y + .1477621 * ( FN
     F(D + C) +  FN F(D - C)))
3470  RETURN
```

This program required less than 1 second to complete on an Apple II computer. The output is as follows:

```
INTEGRAL=2.00000009
```

7.8 CONSIDERATIONS IN THE SELECTION OF A METHOD FOR NUMERICAL DIFFERENTIATION OR INTEGRATION

The selection of an appropriate algorithm for a given differentiation or integration task problem will depend strongly on the nature of the data being manipulated. When selecting an algorithm for use on the microcomputer, the overall limitations of this device must also be kept in mind.

For numerical differentiation, great care should be exercised when using experimental data since the presence of noise can often give results that have little or no meaning. The use of Lagrangian interpolation formulas can help with the differentiation of experimental data as long as one does not try to gain more from the interpolation polynomial than is present. For example, a second-order interpolation polynomial will not give much information about higher-order derivatives. To achieve approximations to higher-order derivatives, it is best to use higher-order interpolation polynomials rather than use a lower-order form in a repeated fashion. When using interpolation polynomials for differentiation, it is well to remember that central difference forms give better results than forward or backward difference forms.

For numerical integration, if the function to be integrated is given in the form of experimental data points, the method of Romberg and Gaussian quadrature can generally not be employed. For this situation the Newton–Cotes formulas must be used. Although these formulas generally require equally spaced data values, special subroutines can be prepared to account for unequal spacing. (This procedure is left as an example problem at the end of this text.) When the function to be integrated can be determined, either the Romberg integration method or Gaussian quadrature may be used. The advantage to Gaussian quadrature is that the user knows in advance how many functional evaluations will be required to achieve an answer and must change methods if the accuracy is not sufficient. With the Romberg method, the user is permitted to specify the desired degree of accuracy at the start, but must be willing to carry out as many functional evaluations as are necessary to get convergence. A concern about the number of functional evaluations can be important for situations involving large computational efforts to achieve function values.

REFERENCES

1. ABRAMOWITZ, MILTON, and IRENE A. STEGUN, eds., *Handbook of Mathematical Functions with Formulas, Graphs and Mathematical Tables*, U.S. Government Printing Office, Washington, D.C., 1964.
2. BICKLEY, W.G., "Formulae for Numerical Differentiation," *Mathematical Gazette*, vol. 25, 1941, 19-27.
3. BURDEN, RICHARD L., J. DOUGLAS FAIRES, and ALBERT C. REYNOLDS, *Numerical Analysis*, Prindle, Weber & Schmidt, Inc., Boston, 1978.
4. DAVIS, PHILIP J., and PHILIP RABINOWITZ, *Numerical Integration*, Blaisdell Publishing Company, Waltham, Mass., 1967.
5. DAVIS, PHILIP J., and PHILIP RABINOWITZ, *Methods of Numerical Integration*, Academic Press, Inc., New York, 1975.
6. EVANS, D.J., *Software for Numerical Mathematics*, Academic Press, Inc., New York, 1974.
7. MACON, NATHANIEL, *Numerical Analysis*, John Wiley & Sons, Inc., New York, 1963.
8. McCALLA, THOMAS R., *Introduction to Numerical Methods and FORTRAN Programming*, John Wiley & Sons, Inc., New York, 1967.
9. PALL, GABRIEL A., *Introduction to Scientific Computing*, Appleton-Century-Crofts Meredith Corp., New York, 1971.
10. RALSTON, ANTHONY, *A First Course in Numerical Analysis*, McGraw-Hill Book Co., New York, 1965.

11. SMITH, W. ALLEN, *Elementary Numerical Analysis*, Harper & Row, Publishers, Inc., New York, 1979.

12. STROUD, A.H., and DON SECREST, *Gaussian Quadrature Formulas*, Prentice-Hall, Inc., Englewood Cliffs, N.J., 1966.

13. VANDERGRAFT, JAMES S., *Introduction to Numerical Computations*, Academic Press, Inc., New York, 1978.

14. WILLIAMS, P.W., *Numerical Computation*, Harper & Row, Publishers, Inc., New York, 1972.

Glossary of computer terms

A

ACCESS TIME. The time required to read data to or from an input or output device.

ACCUMULATOR. A register that stores the results of a computer operation.

ADDER. A logic circuit that adds binary numbers.

ADDRESS. A binary number that identifies a specific memory storage location.

ALGORITHM. A computational procedure for solving a problem.

ALPHANUMERIC. A collection of characters containing both letters of the alphabet and numbers.

ANALOG-TO-DIGITAL. A device that converts external signals to a form the computer can recognize.

APL. **A P**rogramming **L**anguage. A higher-level terminal-oriented programming language.

APPLICATIONS PROGRAM. A program that solves a specific problem, such as inventory control or machine control.

ARITHMETIC-LOGIC UNIT. A logic circuit that performs both arithmetic and logical operations in a digital computer.

ARRAY. A list or table (matrix) of data.

ASCII CODE. An acronym for American Standard Code for Information Interchange. A binary code that represents alphanumeric characters and various symbols.

ASSEMBLER. A computer program that automatically converts assembly language mnemonics into machine language.

ASSEMBLY LANGUAGE. The next step above machine language. Substitutes easily remembered mnemonics for binary machine language instructions.

ASYNCHRONOUS. A computer operation that takes place whenever input information appears. The basic RS flip-flop, for example, is an asynchronous circuit.

AUXILIARY STORAGE. A storage that supplements the internal storage of the central processing unit. Also called secondary storage.

BACKGROUND PROCESSING. Processing of a low-priority program that takes place only when no higher-priority or real-time processing function is present.

BASE. The radix of a number system.

BASIC. An acronym for **B**eginner's **A**ll-Purpose **S**ymbolic **I**nstruction **C**ode. A computer language used with most personal computers.

BATCH PROCESSING. A technique by which items to be processed must be coded and collected into groups prior to processing.

BINARY. A number system with the base 2. Also, a term used to describe a condition or electronic circuit that has only two states, usually on or off.

BINARY CODED DECIMAL (BCD). A number system used in digital computers and calculators that assigns a binary number to each of the ten decimal digits.

BINARY DIGIT. The binary digits 0 or 1.

BISTABLE. An electronic circuit or device that has two operating states, such as a mechanical switch, indicator lamp, or flip-flop.

BIT. An abbreviation for binary digit.

BIT RATE. The rate at which binary digits, or pulse representations, appear on communications lines or channels.

BLOCK. A set of things, such as digits, characters, or words, handled as a unit.

BRANCH. A computer program procedure that transfers control from one instruction to another instruction elsewhere in the program.

BUFFER. A circuit that isolates one circuit from another circuit.

BUG. An error. It can be a mistake in a computer program or a defect in the operation of a computer.

BUS. One or more electrical conductors that transmit power or binary data to the various sections of a computer.

BYTE. A group of (usually) eight binary bits.

CALCULATOR. A microprocessor-based instrument designed primarily for solving mathematical problems.

CARD READER. A computer input mechanism that reads out the information contained on a punched card.

CARRIAGE. A control mechanism for a typewriter or printer that automatically feeds, skips, spaces, or ejects paper forms.

CASSETTE UNIT. A magnetic-tape recorder that uses cassette tapes for storage. Widely used with microcomputers.

CENTRAL PROCESSING UNIT (CPU). The arithmetic-logic unit and control sections of a digital computer.

CHARACTER. Any letter, number, or symbol that a digital computer can understand, store, or process.

CHIP. A thin slice of silicon up to a few tenths of an inch square with an integrated circuit containing from dozens to thousands of electronic parts on its surface.

CIRCUIT. A collection of electronic parts and electrical conductors that performs some useful operation.

CLOCK. A circuit that produces a sequence of regularly spaced electrical pulses to synchronize the operation of the various circuits in a digital computer.

COBOL. **CO**mmon **B**usiness-**O**riented **L**anguage. A higher-level programming language developed for programming business problems.

CODE. A method of representing letters, numbers, symbols, and data with binary numbers.

CODING. The process of translating problem logic represented by a flowchart into computer instructions and data.

CODING FORM. A form on which the instruction for programming a computer are written. Also called a coding sheet.

COMPATIBLE. A term applied to a computer system which implies that it is capable of handling programs devised for some other type of computer system.

COMPILER. A computer program that translates a high-level source-language program into machine-language programs suitable for execution on a particular computing system.

COMPUTER. An electronic device that processes discrete (digital) or approximate (analog) data.

COMPUTER SCIENCE. The field of knowledge that involves the design and use of computer equipment, including software development.

COMPUTER SYSTEM. A central processing unit together with one or more peripheral devices.

CONSOLE. That part of a computer used for communications between the computer operator or maintenance engineer and the computer.

CONTROL PANEL. A panel with input switches and output indicators that allows control of a computer system.

CONTROL SECTION. The electronic nerve center of a digital computer, the circuits that decode incoming instructions and activate the various sections of the computer in perfect synchronization. Part of the Central Processing Unit (CPU).

CONTROL UNIT. The portion of the central processing unit that directs the step-by-step operation of the entire computing system.

CONVERSATIONAL MODE. A mode of operation where a user is in direct contact with a computer and interaction is possible between human and machine without the user being conscious of any language or communications barrier.

COUNTER. A string of flip-flops that counts in binary.

CPU. The Central Processing Unit.

CRT. An acronym for Cathode-Ray Tube. The video display tube used in television sets and many computer terminals.

CYCLE. A specific time interval during which a regular sequence of computer events takes place.

DATA. Numbers, facts, information, results, signals, and almost anything else that can be fed into and processed by a computer.

DATA BASE. A comprehensive data file containing information in a format applicable to a user's needs and available when needed.

DATA COLLECTION. The gathering of source data to be entered into a computer system.

DATA PROCESSING. A term used in reference to operations performed by data-processing equipment.

DATA-PROCESSING CENTER. An installation of computer equipment that provides computing services for users.

DATA STRUCTURES. Arrangement of data (i.e., arrays, files, etc.).

DEBUG. The process of finding and fixing an error in a computer program or in the actual design of a computer.

DECIMAL. A number system with the base 10.

DECODER. A combinational circuit that converts binary data into some other number system.

DECREMENT. To decrease the value of a number by some fixed value, often 1.

DEMULTIPLEXER. A combinational circuit that applies the logic state of a single input to one of several outputs.

DIAGNOSTICS. Statements printed by an assembler or compiler indicating mistakes detected in a source program.

DIGIT. A character in a number system that represents a specific quantity.

DIGITAL. Pertaining to the utilization of discrete integer numbers in a given base to represent all the quantities that occur in a problem or a calculation.

DIGITAL COMPUTER. A computer that uses discrete signals to represent numerical quantities. Nearly all modern digital computers are two-state, binary machines.

DIGITAL PLOTTER. An output unit that graphs data by an automatically controlled pen, plotted as a series of incremental steps.

DIGITAL-TO-ANALOG. A device that converts computer digital data into analog signals.

DIRECT-ACCESS STORAGE. Pertaining to the process of obtaining data from or placing data into storage where the time required for such access is independent of the location of the data most recently obtained or placed in storage. Also called random-access storage.

DISKETTE. See *floppy disk.*

DISK MEMORY. See *magnetic disk memory.*

DISK OPERATING SYSTEM. Software used to manage disk files and programs and to develop application software. Abbreviated DOS.

DISK PACK. The vertical stacking of a series of magnetic disks in a removable, self-contained unit.

DISPLAY UNIT. A device that provides a visual representation of data. See *CRT.*

DOCUMENTATION. An important part of computer design and program development. The process of recording in organized format a detailed list of operational or programming considerations.

DOWN TIME. The total elapsed time that a computer system is unusable because of a malfunction.

DUMP. Printing all or part of the contents of a storage device.

EBCDIC. **E**xtended **B**inary **C**oded **D**ecimal **I**nterchange **C**ode. An 8-bit code used for data representation in several computers.

EMULATE. To imitate one system with another such that the imitating system accepts the same data, executes the same programs, and achieves the same results as the imitated system.

ENCODER. A combinational circuit that converts data from some other number system into binary.

EPROM. Erasable Programmable Read-Only Memory. A read-only storage device that can be erased to change its contents.

ERASE. To clear or remove data from a memory.

EXECUTE. To comply with or act upon an instruction in a digital computer program.

FIELD. A particular category or grouping of data or instructions.

FILE. An organized collection of related data.

FIXED WORD. The condition in which a machine word always contains a fixed number of bits, characters, bytes, or digits.

FLIP-FLOP. The basic sequential logic circuit. A circuit that is always in one of two possible states.

FLOPPY DISK. A flexible disk (diskette) of oxide-coated Mylar that is stored in a plastic envelope. The entire envelope is inserted in the disk unit. Floppy disks are low-cost storage units and are widely used with microcomputers and minicomputers.

FLOW CHART. A diagram that shows the major steps or operations that take place in a computer program.

FOREGROUND PROCESSING. The automatic execution of the computer programs that have been designed to preempt the use of the computing facilities.

FORTRAN. **FOR**mula **TRAN**slation. A higher-level programming language designed for programming scientific-type problems.

GARBAGE. A term often used to describe incorrect answers from a computer program, usually resulting from equipment malfunction or a mistake in a computer program.

GATE. The simplest electronic logic circuit. A single gate may invert the logic state at its input or make a simple decision about the status of two or more inputs.

GENERAL-PURPOSE COMPUTER. A computer that is designed to solve a wide class of problems. The majority of digital computers are of this type.

GIGO. Garbage In–Garbage Out. If the input data are poor (garbage in), then the output data will also be poor (garbage out).

HARD COPY. A paper printout of computer results or data.

HARDWARE. The electronic circuits in a computer.

HEXADECIMAL. A number system with the base 16. "Hex" numbers are convenient for representing 4-bit binary groups.

HIGHER-LEVEL LANGUAGE. A computer programming language that is intended to be independent of a particular computer.

HOUSEKEEPING. Operations that take place in a computer or a computer program that clears memories, checks status registers, organizes data, and otherwise sets things up in preparation for a data-processing operation.

ILLEGAL OPERATION. A program instruction that a computer cannot perform.

INCREMENT. To increase the value of a number by some fixed value, often 1.

INFORMATION. Data that have been organized into a meaningful sequence.

INFORMATION RETRIEVAL. A technique of classifying and indexing useful data in mass storage devices in a format amenable to interaction with the user(s).

INPUT. The introduction of data from an external source into the internal storage unit of a computer.

INPUT/OUTPUT. A general term for the peripheral devices used to communicate with a digital computer and the data involved in the communication.

INPUT UNIT. A device used to transmit data into a central processing unit.

INSTRUCTION. A set of characters used to define a basic operation and to tell the computer where to find the data needed to carry out an operation.

INSTRUCTION REPERTOIRE. The complete set of machine instructions for a computer.

INTEGER. A whole number, which may be positive, negative, or zero. It does not have a fractional part. Examples of integers are 526, -378, or 0.

INTEGRATED CIRCUIT. An electronic circuit formed on the surface of a tiny silicon chip.

INTELLIGENT TERMINAL. An input/output device in which a number of computer processing characteristics are physically built into the terminal unit.

INTERACTIVE. Highly communicative between the user and the computer system.

INTERFACE. A common boundary between two pieces of hardware or between two systems.

INTERNAL STORAGE. Addressable storage directly controlled by the central processing unit of a digital computer.

INTERPRETER. A computer program that translates and then executes a computer program a step at a time.

INTERRUPT. To temporarily disrupt the normal execution of a program by a special signal.

ITERATIVE PROCESS. A process in which the same procedure is repeated many times until the desired answer is produced.

JOB. A specified group of tasks prescribed as a unit of work for a computer.

K. A shorthand way of expressing the capacity of a computer memory. Corresponds to 2^{10} (1024). Therefore, a 4K memory stores 4,096 bits.

KEYBOARD. A typewriterlike array of switches used to feed data into a digital computer manually.

LANGUAGE. The symbols, phrases, characters, and numbers used to communicate with a digital computer.

LIGHT PEN. A stylus used with *crt* display devices to add, modify, and delete information on the face of the screen.

LINE PRINTER. A printer that prints a complete line of type in one operation.

LOGIC CIRCUIT. A gate or other circuit that responds to two-state signals.

LOOP. A sequence of computer instructions that is repeated one or more times until a desired result is achieved.

LSI. Large-Scale Integration. Logic used in microprocessors, microcomputers, and other computers.

MACHINE ADDRESS. An address that is permanently assigned by the machine designer to a storage location. Also called absolute address.

MACHINE INDEPENDENT. A term used to indicate that a program is developed in terms of the problem rather than in terms of the characteristics of the computer system.

MACHINE LANGUAGE. Fundamental language of a computer. Programs written in machine language require no further interpretation.

MACROINSTRUCTION. A computer instruction composed of a sequence of microinstructions.

MAGNETIC CORE. A tiny ring of material that can store a single binary bit.

MAGNETIC DISK MEMORY. A memory system that stores and retrieves binary data on recordlike metal or plastic disks coated with a magnetic material.

MAGNETIC TAPE MEMORY. A memory system that stores and retrieves binary data on magnetic recording tape.

MAGNETIC TAPE UNIT. A device used to read and write data in the form of magnetic spots on reels of tape coated with a magnetizable material.

MAINFRAME. The part of the computer that contains the arithmetic unit, internal storage unit, and control functions.

MAIN STORAGE. See *internal storage.*

MATRIX PRINTING. A method of printing characters and other data by a pattern of small dots.

MEDIUM. The physical substance upon which data are recorded (e.g., diskette, magnetic tape).

MAGAHERTZ. Millions of cycles per second.

MEMORY. The part of a digital computer that stores data.

MENU. A display of selections that may be chosen, typically on a visual display screen.

MICROCOMPUTER. A digital computer made by combining microprocessor with one or more memory circuits. Single-chip microcomputers are also available.

MICROELECTRONICS. The field that deals with techniques for producing miniature circuits (e.g., integrated circuits, thin-film techniques, and solid-logic modules).

MICROINSTRUCTION. The most basic operation that takes place in a digital computer.

MICROPROCESSOR. The complete central processing unit for a digital computer (arithmetic-logic unit, control section and some registers) on a single silicon chip.

MICROSECOND. One millionth of a second. Abbreviated μs.

MILLISECOND. One thousandth of a second. Abbreviated ms.

MINICOMPUTER. A small and relatively inexpensive digital computer.

MISTAKE. A human failing (e.g., faulty arithmetic, incorrect keypunching, incorrect formula, or incorrect computer instructions).

MNEMONIC. A memory aid such as an abbreviation or acronym.

MODEM. A contraction of MOdulator/DEModulator. Its function is to interface with data-processing devices and covert data to a form compatible for sending and receiving on transmission facilities.

MONITOR. A television receiver without the circuitry to detect transmissions. Used widely with microcomputers.

MULTIPLEXER. A combinational circuit that applies the logic state of one of several inputs to a single output.

MULTIPROCESSING. Independent and simultaneous processing accomplished by a computer configuration consisting of more than one arithmetic and logic unit, each being capable of accessing a common memory.

MULTIPROGRAMMING. Pertaining to the concurrent execution of two or more programs by a computer. The programs operate in an interleaved manner within one computer system.

NANOSECOND. One billionth of a second. Abbreviated ns.

NEGATIVE LOGIC. A logic system where the binary bit 0 is represented by a high voltage level and the bit 1 by a low voltage level.

NETWORK. The interconnection of a number of points by data communications facilities.

NONVOLATILE STORAGE. A memory system that retains data without the need for electrical power.

NUMBER. The representation of a quantity. In digital computers, numbers can represent data, characters, instructions, and so on.

NUMERIC PAD. A special cluster of keys allowing input of the numeric digits 0 through 9.

OBJECT PROGRAM. A program written in or expressed in machine language.

OCTAL. A numbering system using base 8.

OFF-LINE. Peripheral units that operate independently of the central processing unit.

ON-LINE. Peripheral devices operating under the direct control of the central processing unit.

OPERATING SYSTEM. Software that controls the execution of computer programs and that may provide scheduling, input/output control, compilation, data management, debugging, storage assignment, accounting, and other similar functions.

OPERATION CODE. The portion of an instruction that designates the operation to be performed by a computer (e.g., add, subtract, or move). Also called a command.

OUTPUT. Data transferred from the internal storage unit of a computer to some storage or peripheral device.

OUTPUT SECTION. A printer, video display, or other device that makes information processed by computer available to an operator or an electronic device.

OUTPUT UNIT. A device capable of recording data coming from the internal storage unit of a computer (e.g., card punch, line printer, *crt* display, magnetic disk, or teletypewriter).

PARALLEL PROCESSING. Operating on data a chunk of bits at a time.

PARITY BIT. A binary bit added to a binary word to make the total number of 1's either even or odd.

PASCAL. A programming language that is of particular interest to computer scientists and is being used increasingly for many applications.

PATCH. To temporarily modify the software or hardware of a computer system.

PERIPHERAL DEVICES. Input/output and storage devices attached to a computer.

PERSONAL COMPUTER. A microcomputer with a keyboard input designed for ease of use and maximum economy.

PICOSECOND. One trillionth of a second. Abbreviated ps.

PL/1. Programming Language/1. A general-purpose programming language.

POSITIVE LOGIC. A logic system where the binary bit 1 is represented by a high voltage level and the bit 0 by a low voltage level.

PRECISION. The degree of exactness with which a quantity is stated. The result of a calculation may have more precision than it has accuracy. For example, the true value of pi to six significant digits is 3.14159; the value 3.14162 is precise to six digits, given to six digits, but is accurate to only about five.

PRINTER. An output device that prints computer information on paper.

PROBLEM-ORIENTED LANGUAGE. A programming language designed for the convenient expression of a given class of problems (e.g., GPSS).

PROCEDURE. A precise step-by-step method for effecting a solution to a problem.

PROCEDURE-ORIENTED LANGUAGE. A programming language designed for the convenient expression of procedures used in the solution of a wide class of problems (e.g., FORTRAN, APL, PL/1, BASIC, and Pascal).

PROCESSOR. A digital computer.

PROGRAM. A list of instructions that tells a computer what to do and how to do it.

PROGRAMMING. The process of translating a problem from its physical environment to a language that a computer can understand and obey.

PROGRAMMING LANGUAGE. A language used to prepare computer programs.

PROM. Programmable Read-Only Memory. A read-only memory, programmable by the purchaser, that cannot be erased.

PROMPT. A character(s) printed by the program to signal the user that input is required.

RAM. Read/Write Memory.

RANDOM-ACCESS MEMORY. A memory that offers equal access time to any storage location.

RAW DATA. Data that have not been processed.

READ. To sense data from a magnetic tape, disk, or punched card. Or to make information in a memory available to some other circuit.

READER. Any device capable of transcribing data from an input medium.

READ-ONLY MEMORY. A memory that contains permanent data that cannot be altered or erased. Usually designated ROM.

READ/WRITE MEMORY. A memory that contains information that can be erased and modified. Often designated RAM.

REAL-TIME SYSTEM. A system where transactions are processed as they occur.

RECORD. A group of related items of data treated as a unit (e.g., the inventory master record). A complete set of such records forms a file.

REGISTER. A string of flip-flops that stores one word of binary data. A register is a temporary memory.

RELIABILITY. A measure of the ability to function without failure.

RELOCATE. To move a routine from one portion of storage to another and to adjust the necessary address references so that the routine, in its new location, can be executed.

REMOTE PROCESSING. The processing of computer programs through an input/output device that is remotely connected to a computer system.

REMOTE TERMINAL. An input/output device that is remotely located from a computer system. Also called remote station.

RESOLUTION. The quality of a visual display.

RESPONSE TIME. The time it takes the program or input/output device to respond to a user input or command.

ROM. Read-Only Memory. A type of memory permanently programmed by the manufacturer.

RUN. A single, continuous performance of a computer program.

SCAN. To retrieve or store data from beginning to end of a list or table.

SCREEN. The picture tube of a visual display terminal.

SCROLL. To display various sections of a long list of lines, similar to viewing a portion of a scroll as it is unwound.

SEMICONDUCTOR MEMORY. A computer memory that uses silicon integrated-circuit chips.

SEQUENTIAL LOGIC. A collection of logic gates that responds to incoming information only when a clock pulse is received. Sequential logic circuits use flip-flops so that each operation is affected by a previous operation.

SERIAL PROCESSING. Operating on data a bit at a time.

SIMULATE. To represent the functioning of one system by another (i.e., to represent a physical system by the execution of a computer program).

SOFTWARE. Paperwork such as programs and documentation associated with the operation of a computer.

SOLID STATE. The electronic components that convey or control electrons within solid materials (e.g., transistors, germanium diodes, and integrated circuits).

SORT. To arrange numeric or alphabetic data in a given sequence.

SOURCE COMPUTER. A computer used to translate a source program into an object program.

SOURCE DOCUMENT. An original document from which basic data are extracted (e.g., invoice, parts list, inventory tag).

SOURCE LANGUAGE. The orginal form in which a program is prepared prior to processing by the computer (e.g., FORTRAN or assembly language).

SOURCE PROGRAM. A computer program written in a nonbinary form such as assembly language or BASIC.

SPECIAL CHARACTER. A graphic character that is neither a letter nor a digit (e.g., the plus sign and the period).

SPECIAL-PURPOSE COMPUTER. A computer designed to solve a specific class or narrow range of problems.

STANDARD. An accepted and approved criterion used for writing computer programs, drawing flow charts, building computers, and so on.

STATEMENT. The most elemental instruction to the computer in a higher-level programming language, such as BASIC or FORTRAN.

STORAGE. The retention of data so that they can be obtained at a later time.

STORAGE DEVICE. A computer memory.

STORAGE LOCATION. A position in storage where a character, byte, or word may be stored.

STORAGE MAP. An aid used by the computer user for estimating the proportion of storage capacity to be allocated to data and instructions.

STORAGE PROTECTION. A device that prevents a computer program from destroying or writing in computer storage beyond certain boundary limits.

STORAGE UNIT. The portion of the central processing unit that is used to store instructions and data.

STRUCTURED PROGRAMMING. Techniques concerned with improving the programming process through better organization of programs and better programming notation to facilitate correct and clear description of data and control structures.

SUBROUTINE. A sequence of instructions in a computer program that is used more than once by the main program.

SYMBOLIC ADDRESS. An address expressed in symbols.

SYMBOLIC CODING. Coding in which the instructions are written in non-machine language (e.g., a FORTRAN program).

SYNCHRONOUS. A computer operation that takes place under the control of a clock.

SYSTEM. An organized collection of machines, methods, and personnel required to accomplish a specific objective.

SYSTEMS ANALYSIS. The examination of an activity, procedure, method, technique, or a business to determine what must be accomplished and how the necessary operations may best be accomplished by using electronic data-processing equipment.

SYSTEMS ANALYST. A person skilled in solving problems with a digital computer. He or she analyzes and develops information systems.

SYSTEMS PROGRAMS. Computer programs provided by a computer manufacturer. Examples are operating systems, assemblers, compilers, debugging aids, and input/output programs.

SYSTEMS STUDY. The detailed process of determining a set of procedures for using a computer for definite operations, and establishing specifications to be used as a basis for the selection of equipment suitable to the specific needs.

TELECOMMUNICATIONS. Pertaining to the transmission of data over long distances through telephone and satellite facilities.

TELETYPEWRITER. A typewriterlike device that can be used to feed data and programs into a computer and to print the output information from a computer on a strip of paper.

TERMINAL. A computer input/output device.

TEXT EDITING. The capability to modify text automatically.

THIRD GENERATION. Computers made with integrated circuits.

TIME SHARING. A method of operation whereby a computer system automatically distributes processing time among many users simultaneously.

TRACK. The path along which data are recorded, as in magnetic disks and magnetic drums.

TRANSLATE. To change data from one form of representation to another without significantly affecting the meaning.

TRANSLATION. Conversion of a higher-level language, or assembly language, to machine-understandable form. See *assembler, compiler*, and *interpreter*.

TURNKEY SYSTEM. A computer system that includes all hardware and software to perform a specified application without the need of professional computer personnel.

TYPEWRITER. An input/output device that is capable of being connected to a digital computer and used for communications purposes.

VARIABLE. A quantity that can assume any of a given set of values.

VIDEO. Signals that generate display of data on a visual display terminal.

VIRTUAL MEMORY. A technique for managing a limited amount of high-speed memory and a (generally) much larger amount of lower-speed memory in such a way that the distinction is largely transparent to a computer user.

VOLATILE STORAGE. A memory system that retains data only when electrical power is present.

WORD. A string of binary bits used to represent a number, character, or instruction in a digital computer. Computer words can be any length.

WORD LENGTH. The number of bits, bytes, or characters in a word.

WORD PROCESSING. A text-editing system. A system that is used to prepare text, such as letters, forms, and pages of text.

WRITE. To place information into a memory or register.

Hex (ASCII table)

B

Control		*Numeric*		*Uppercase*		*Lowercase*		*Special*	
00	Null	30	0	41	A	61	a	20	Space
01	Start of Heading	31	1	42	B	62	b	21	!
02	Start of Text	32	2	43	C	63	c	22	"
03	End of Text	33	3	44	D	64	d	23	#
04	End of Transmission	34	4	45	E	65	e	24	$
05	Enquiry	35	5	46	F	66	f	25	%
06	Acknowledge	36	6	47	G	67	g	26	&
07	Bell	37	7	48	H	68	h	27	'
08	Backspace	38	8	49	I	69	i	28	(
09	Horizontal Tabulation	39	9	4A	J	6A	j	29	)
0A	Line Feed			4B	K	6B	k	2A	*
0B	Vertical Tabulation			4C	L	6C	l	2B	+
0C	Form Feed			4D	M	6D	m	2C	,
0D	Carriage Return			4E	N	6E	n	2D	-
0E	Shift Out			4F	O	6F	o	2E	.
0F	Shift In			50	P	70	p	2F	/
10	Data Link Escape			51	Q	71	q	3A	:
11	Device Control 1			52	R	72	r	3B	;
12	Device Control 2			53	S	73	s	3C	<
13	Device Control 3			54	T	74	t	3D	=
14	Device Control 4			55	U	75	u	3E	>
15	Negative Acknowledge			56	V	76	v	3F	?
16	Synchronous Idle			57	W	77	w	40	@
17	End of Transmission Block			58	X	78	x	5B	[
18	Cancel			59	Y	79	y	5C	\
19	End of Medium			5A	Z	7A	z	5D	]
1A	Substitute							5E	^ or ↑
1B	Escape							5F	_ or ←
1C	File Separator							7B	{
1D	Group Separator							7C	|
1E	Record Separator							7D	}
1F	Unit Separator							7E	~
7F	Rub Out or Delete								

Time units

C

Time Units and Their Fractional Equivalents

Time Unit	*Notation*	*Fractional Equivalent*	*Abbreviation*
1 second	10^{0}	1 second	sec
1 millisecond	10^{-3}	1/1000 second	ms
1 microsecond	10^{-6}	1/1,000,000 second	μs
1 nanosecond	10^{-9}	1/1,000,000,000 second	ns
1 picosecond	10^{-12}	1/1,000,000,000,000 second	ps

Number conversion techniques

D

CONVERSION OF NUMBERS INTO THE BASE 10 SYSTEM

To understand the theory of number conversion from one base to another, it will be useful to look at the meaning of a number in the decimal system. Each decimal digit in a number represents a coefficient to be multiplied by the appropriate power of 10, the base. The position of the coefficient in the number string indicates which power of 10 is to be used in generating the products. The value of the number will be the sum of these products. Thus the number 125.64_{10} means

$$1. \times 10^2 + 2. \times 10^1 + 5. \times 10^0 + 6. \times 10^{-1} + 4. \times 10^{-2}$$

Any number in any base can be converted to its base 10 equivalent by summing the appropriate powers of the base, each multiplied by the proper coefficient. For example, 241.56_8 means

$$\begin{aligned} 2. \times 8^2 &= 128. \\ +4. \times 8^1 &= 32. \\ +1. \times 8^0 &= 1. \\ +5. \times 8^{-1} &= 0.625 \\ +6. \times 8^{-2} &= 0.093\ 75 \\ \hline &= 161.718\ 75_{10} \end{aligned}$$

or $C4.A_{16}$ means

$$\begin{aligned} C(12.) \times 16^1 &= 192. \\ +4. \times 16^0 &= 4. \\ +A(10.) \times 16^{-1} &= 0.625 \\ \hline & 196.625_{10} \end{aligned}$$

or 110.101_2 means

$$\begin{aligned} 1 \times 2^2 &= 4. \\ +1 \times 2^1 &= 2. \\ +0 \times 2^0 &= 0. \\ +1 \times 2^{-1} &= 0.5 \\ +0 \times 2^{-2} &= 0. \\ +1 \times 2^{-3} &= 0.125 \\ \hline & 6.625_{10} \end{aligned}$$

CONVERSION OF NUMBERS FROM BASE 10 INTO OTHER BASES

Conversion from base 10 into other new bases requires a two-part procedure. The first part of the procedure is used to convert the digits to the left of the radix point (i.e., the whole number part). This is done by first dividing the base 10 number by the new base. This division may be accomplished using base 10 mathematics. The division process will yield an answer in the form of dividend and a remainder. The remainder becomes the first converted digit to the left of the radix point. Next, the dividend is divided by the new base to yield a second dividend and a new remainder, which becomes the second converted digit. This process is repeated until the dividend is small enough to constitute only a remainder to become the final converted digit.

The conversion of 74_{10} to base 8 would consist of

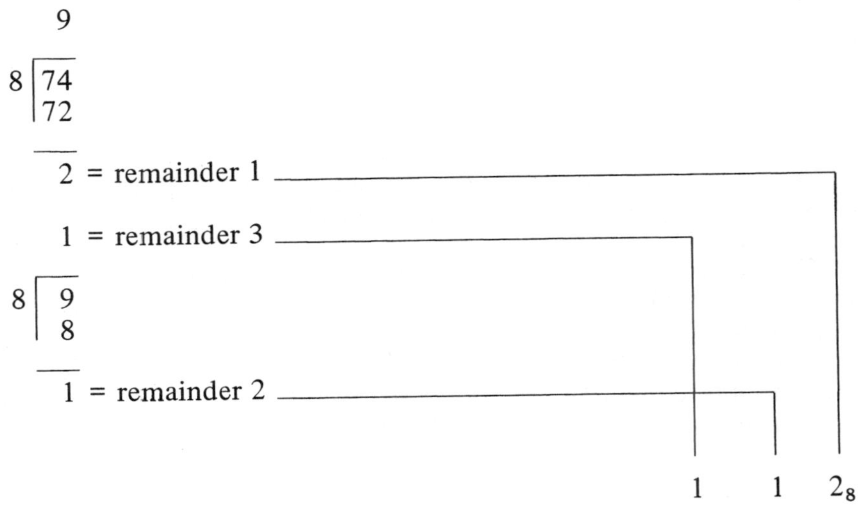

The second part of the conversion procedure is used to convert all the digits to the right of the decimal point (i.e., the fractional part). This conversion is done by multiplying the fractional part by the new base. The resulting product will be an integer plus a new fractional part. The integer is the first converted digit to the right of the decimal point, and the leftover fraction is used to generate the second converted digit by multiplying again by the new base. The result will be another integer plus a fractional part. This second integer is the second converted digit to the right of the decimal, and this second, leftover fraction is used to generate the third converted digit. The procedure continues until the leftover fraction becomes zero or until the desired number of significant digits is achieved.

For example, 0.55_{10} converted to base 8 will be

$8 \times 0.55 = 4.40$	First digit = 4
$8 \times 0.40 = 3.20$	Second digit = 3
$8 \times 0.20 = 1.60$	Third digit = 1
$8 \times 0.60 = 4.80$	Fourth digit = 4
$8 \times 0.80 = 6.40$	Fifth digit = 6
$8 \times 0.40 = 3.20$	Sixth digit = 3
$\vdots$	

so that

$$0.55_{10} = 0.4\underbrace{3146}\underbrace{3146}\underbrace{3146}\ldots_8$$

repeating pattern

In this example we note that a repeating pattern appears. It is not always possible to find an exact equivalent for a fraction in two different base systems.

The main storage for a digital computer can only store a finite number of digits for a given fraction. It should also be mentioned that the digital computer does not round off answers in its printout. Rather, it is designed simply to truncate the values. Thus, if the number 0.55_{10} is given to the computer, converted to base 2, stored, retrieved from storage, and reconverted to base 10, the result may be truncated and printed as 0.549999_{10}.

RS-232C interface connections

E

Pin Number	*Definition*	*Signal Direction*
1	Ground	–
2	Transmit data	To DCE
3	Receive data	From DCE
4	Request to send	To DCE
5	Clear to send	From DCE
6	Data set ready	From DCE
7	Ground	–
8	Data carrier detect	From DCE
20	Data terminal ready	To DCE

Index

A

B

C

D

E

F

G

H

I

J

K

L

M

N

O

P

Q

R

S

T

V

W